Mathematics for Industry

Volume 12

Aims & Scope

The meaning of “Mathematics for Industry” (sometimes abbreviated as MI or MfI) is different from that of “Mathematics in Industry” (or of “Industrial Mathematics”). The latter is restrictive: it tends to be identified with the actual mathematics that specifically arises in the daily management and operation of manufacturing. The former, however, denotes a new research field in mathematics that may serve as a foundation for creating future technologies. This concept was born from the integration and reorganization of pure and applied mathematics in the present day into a fluid and versatile form capable of stimulating awareness of the importance of mathematics in industry, as well as responding to the needs of industrial technologies. The history of this integration and reorganization indicates that this basic idea will someday find increasing utility. Mathematics can be a key technology in modern society.
The series aims to promote this trend by (1) providing comprehensive content on applications of mathematics, especially to industry technologies via various types of scientific research, (2) introducing basic, useful, necessary and crucial knowledge for several applications through concrete subjects, and (3) introducing new research results and developments for applications of mathematics in the real world. These points may provide the basis for opening a new mathematics-oriented technological world and even new research fields of mathematics.

More information about this series at http://www.springer.com/series/13254

Yoichi Ando

Opera House Acoustics Based on Subjective Preference Theory

Springer

Yoichi Ando
Kobe University
Kobe
Japan

ISSN 2198-350X ISSN 2198-3518 (electronic)
Mathematics for Industry
ISBN 978-4-431-55422-6 ISBN 978-4-431-55423-3 (eBook)
DOI 10.1007/978-4-431-55423-3

Library of Congress Control Number: 2015932241

Springer Tokyo Heidelberg New York Dordrecht London

Printed on acid-free paper

Springer Japan KK is part of Springer Science+Business Media (www.springer.com)

An expression of individual personality, “a seed” of creations for the third stage of human life. Painted by Massimo Cocchi in 1980

Foreword

When Prof. Yoichi Ando invited me to present a paper in a special session of ICA 92 in Beijing, I already knew and appreciated his fundamental work on Concert Hall Acoustics (1985). That meeting was fruitful both on the scientific and on the human relationship points of view and since then he has honored me with his friendship, which has been full of reciprocal invitations in our countries and has everytime left me enriched not only with new notions and concepts of architectural acoustics, Opera Houses in particular, but also about auditory and visual perception. This blend of ecleptic concepts, notions, and methods is the fundamental basis of Temporal Design, the fruit which ripened from his outstanding scientific work.

Following his idea about the organization of human life into three stages, I may note that now both of us are in the third stage and we are trying to leave to the human culture the best of the activity that enriched the second stage of our lives.

My activity as in acoustic engineering developed, among others, in the field of Theater design and my target is to sketch a history of the cavea design starting from the analysis of the acoustical knowledge of the architect that designed it at different moments of human history, from the age of the Ancient Greeks, to the modern age, passing through the period of Renaissance.

Buildings constructed during each period of the human history reflect the top cultural level of the same, so the Greeks built theaters whose shape was suggested only from the knowledge of acoustical waves, frequency, and related wavelength and dimensions were related to the ability to clearly receive the acoustical performance and the visual message connected with it: even the location in the field was determined by the spectators' ability to see and listen and the only way they knew to build tiers was to lay them on a natural declivity (for instance, Cocchi 2013). Aristotle had also some knowledge of reflection of sound, so it is allowed to think that the theaters designed by Aristotelians already took advantage of the acoustical effect of the orchestra area, backstage and lateral walls, and so on, till the Romans, who knowing how to build pillars and arcs, were able also to partially or totally cover the cavea. Many Roman theaters were erected on the writings of Vitruvius (ca 25 BC); about more than a millennium of decline of scientific thought followed that historical period. It was in the Renaissance that the nobility's cultural concern gave new impulse to the construction of theaters, at the beginning inside palaces (like in Vicenza, Sabbioneta, and Parma) and following the writings of Vitruvius, then in open spaces (for instance, The Globe in London), till the Opera Houses of the eighteenth and nineteenth centuries.

The acoustic quality of the last mentioned Opera Houses is usually prised and taken as an example from those who try to pick out the secrets of the architects of that time, but it is quite impossible to find papers from them, so we can only apply to these cavea the modern measuring means and collecting individual judgements (for instance, Beranek 1962 and Barron 1993) and infer from them some designing rule.

Even if someone wrote about the reasons why theaters cavea must be designed according to Vitruvians' ideas (for instance, Milizia 1773–1794), in my opinion, till the first half of the last century the only new idea based on a true scientific approach is that of Sabine (1900) who found the relevant role of reverberation in acoustical perception.

In the computer era, Prof. Ando faced the problem from an original point of view, searching for the neural connection between perception (Chap. 4) and preference (Chaps. 5 and 6), investigating also at the human brain level (Chap. 4), developing a quantitative approach that is indeed free from those subjective conditionings that can influence personal judgements.

His thought and experiences are now scoured and summarized in this book, whose primary merits are, from my point of view, not only to have reorganized the excess of parameters born from the computer era into only four independent ones (Chaps. 2 and 3), but chiefly to have stated a link between temporal and spatial factors, physical parameters easily evaluable in the field (Chaps. 7 and 8), and preference (Chap. 6). Not to be undervaluated the link between the two senses that enables the brain to be in continuous connection with the world, the acoustic and visual ones (Chap. 11). Since Opera as an artistic discipline is a beautiful blend of music and imagery, therefore it is really a great achievement that a scientific theory capable of accounting both factors has been developed.

This book presents also many suggestions useful both for performers (Chap. 9) and designers (Chaps. 10 and 12) that enable to view this work as a bridge between the scientific analysis of the Opera House and those that share the same interest but from a practical point of view.

At first reading, people not already acquainted with the subject may proceed step-by-step, as every concept is clearly exposed but in a synthetic way, being sure that any available source has been deeply analyzed: the very long list of references means that it is very hard to master any concept if has to carefully refer to the suggested papers, otherwise one must trust for sure that any statement has been accurately considered by the author.

January 2015

Professor em. Alessandro Cocchi
University of Bologna

Preface

Based on individual personality, we are all creators and performers.

Since 1985, when the first volume of this series, *Concert Hall Acoustics*, was published, remarkable progress has been made in temporal- and spatial-primary percepts of sound. The subjective preference theory, well based on neural evidence of the sound field, has been developed. Thus, a model of the auditory pathway with brain activities has been reconfirmed (Ando 1985, 1998, 2009). The specialization of the left and right human cerebral hemispheres that support the model of the auditory-brain system has been well described. Neural activities related to subjective preference of the sound field and the visual field have been discovered.

Subjective preference is made up of the most primitive responses of subjective attributes, because preferences are an evaluative judgment, and judgment is performed in the direction of maintaining life and is deeply related to aesthetic issues. Overall, subjective responses including the annoyance of environmental noise, speech recognition (Ando 2015), and reverberance as well as subjective preference of the sound field may be well described by both temporal and spatial factors. Such significant temporal and spatial factors are extracted from the running autocorrelation function and the interaural cross-correlation function, respectively.

A new possible type of opera house can be designed by the maximization of the scale value of subjective preference of the sound field applying the genetic algorithm (GA).

Also, a wide range of applications of this model is available including those for quality of the sound field in an opera house with the stage for vocal sources and the pit for musical instruments, and the visual field on the stage can be well designed.

This volume focuses on *Opera House Acoustics Based on Subjective Preference Theory*. The author aims to present information to researchers and students in acoustics and vision who are interested in physics, psychology, brain physiology, and understanding of any subjective attributes in relation to objective parameters.

The well-known Helmholtz theory, which was based on a peripheral model of auditory system, unfortunately has failed to describe pitch, timbre, and duration as well as spatial sensations, thus it also fails to describe overall responses such as subjective preference of sound fields and annoyance of environmental noise and even speech recognition without a supercomputer.

Acknowledgments

The subjective preference theory was established by a series of investigations since 1975 at the Third Physics Institute of the University of Göttingen, where Director Professor Manfred Schröder sent a recommendation to the Alexander von Humboldt Foundation in Bonn to invite the present author to his institute.

Colleagues of the Yoichi Ando Laboratory, the Graduate School at Kobe University; the Professor Alessandro Cocchi Laboratory, University of Bologna; and the Professor Roberto Pompoli Laboratory, University of Ferrara; provided useful information as well as illustrations for this volume. Particularly, Alessandro Cocchi provided the photograph of a drawing by his father, Massimo Cocchi, which is printed at the very beginning of this volume. Shin-ichi Sato, Hiroyuki Sakai, Nicola Prodi, Yoshiharu Soeta, Kenji Fujii, Ryota Shimokura, Yosuke Okamoto, and Kosuke Kato have published a number of excellent works, which are fully cited in this volume. The author would like to express his appreciation to the laboratories, the authors of papers, and the publishers who have granted permission for the use of their works for publication in this volume. Keiko Ando suggested the term "crystal opera house" as discussed in Sect. 12.2.

Drs. Akira Fujimori and Shioko Okada, Konan Hospital, Kobe, oversaw continuous medical treatments to maintain the health of the author.

Contents

Chapter 1
Introduction

Drama, together with music, is performed usually in an opera house, which is played on the stage and in the pit as an epitome of human lives. The audience may much enjoy, and obtain a hint of their future life for worthwhile lives through personality given by Nature. As is widely known, even newborn babies always bring messages from Nature with a lovely smile that was developed by affection since the beginning of the universe. A new type of opera house cooperating with natural activities is expected for each individual that may play, which means "we are all on the stage." In order to realize creations through individual personality, an opera house is proposed with a sketch in the last chapter of this volume.

The ancient Greek and Roman theaters are the origin of the modern opera houses and drama theaters (Barron 1993). An ancient Greek theater consists of the theatron, consisting of the audience seating and exiting aisles, and the orchestra, a flat acting area. The ancient Roman theater added a large stage building behind the orchestra, patterned after the later Greek (Hellenistic) theaters. This stage building provides to the audience area with strong reflections reinforcing the direct sound, and such reflections improve source loudness and speech intelligibility. In ancient architectural acoustics, the concepts of reverberation, interference, echo disturbance, and clarity of voice were described by Vitruvius (ca 25 BC).

It is often considered that the sound field in the ancient theater consists of direct sound and only a single reflection from the ground and is similar to the sound field of open space (Beranek 1962). The presence of scattered and reverberant sound is assumed negligible. However, it is reported that the unoccupied seats and the heads of the audience also scatter sound to adjacent areas (Shankland 1973).

Since 1985, there has been remarkable progress in temporal- and spatial-primary percepts of sound, and the theory of subjective preference of the sound field based on neural evidences (Ando 1998, 2009a) is the first bridge between science and art. Neural activities have been discovered that correspond well to subjective preferences of the sound field, which are the most primitive responses of subjective attributes, which are the evaluative judgments. This is due to the fact that preferences guide the organism in the direction of maintaining life. In humans, therefore,

Y. Ando, *Opera House Acoustics Based on Subjective Preference Theory*,
Mathematics for Industry 12, DOI 10.1007/978-4-431-55423-3_1

these preferences are deeply related to aesthetic issues. Subjective preference is an overall response in cooperating both temporal and spatial factors associated with the left and right hemispheres. Results of slow vertex responses (SVR) related to preference were discovered in relation to the temporal factors: sensation level and the initial time delay gap between the direct sound and the first reflection. Neural activities of the IACC and listening level (LL) spatial factors associated with the right hemisphere. The activities of temporal factors are associated with the left hemisphere. Any overall subjective responses, including intelligibility of speech and reverberance, as well as subjective preference of the sound field may be well described by both the temporal and spatial factors. The specialization of cerebral hemispheres signifies the independent influence of the temporal and spatial factors on the subjective preference. This supports the model of central auditory signal processing model.

The four temporal primary percepts, namely (1) loudness, (2) pitch, (3) timbre, and (4) duration are well described by the temporal factors extracted from the running autocorrelation function (ACF) of sound signal. Four spatial-primary percepts of the sound field such as (1) localization or direction of sound signal arriving at a listener position, (2) movement of a sound source on the stage (Sect. 2.5.4), (3) apparent source width (ASW), and (4) subjective diffuseness are described by the spatial factors extracted from the running crosscorrelation function (IACF) for the signals arriving at two ears. The significant factors determining these percepts could not be well described by spectrum analyses. Therefore, Helmholtz theory was unable to well describe both temporal percepts such as timbre and pitch as well as spatial percepts.

A wide range of applications of this model is available to measure acoustic parameters and quality of the sound field in an opera house with the stage for vocal sources and the pit for musical instruments, which are quite different from a concert hall. Speech recognition could be well described mainly by the temporal factors (Ando 2015 to be published).

It is suggested that a new possible type of opera house may be designed by the maximization of the scale value of subjective preference. Blending the vocal source on the stage, the orchestra music in the pit and the sound field is of major importance in the opera house acoustics. Integration of measured factors after construction of an opera house including effects of scattered reflections may be utilized for further design studies.

Also, the seamless theory with temporal and spatial factors might be applied for percepts of vision in opera performance on the stage. For example, flickering lights such as for twinkling stars and a movement such as for leaves with air movement as well as texture of walls might be well blended with opera performing on the stage and with musical performance in the pit.

Chapter 2
Analyses of Temporal Factors of a Source Signal

Sound signals proceed along auditory pathways and are perceived in a time sequence while the brain simultaneously interprets the meaning of signals. Thus, a great deal of attention is paid here to analyzing the signal in the time domain. This chapter mainly treats physical aspects of the running autocorrelation function (ACF) of the signal, which contains the envelope and its finer structures as well as the power at its starting time. Mathematically, the ACF has the same information as the power density spectrum of the signal under analysis. From the ACF, however, significant factors may be easily extracted, which are directly related to temporal percepts (such as four temporal primary sensations, i.e., loudness, pitch, timbre and duration are well described by the temporal factors extracted from the running autocorrelation function of sound signal). The ACF processor exists in the auditory pathway not at the very periphery but close to the brain as discussed in Chap. 5, *so that the any psychological responses are affected directly by these factors. And, the running inter-aural crosscorrelation function (IACF) processor exists in the auditory pathway around inferior colliculus. The spatial factors may well describe the spatial percepts (localization or direction of sound signal arriving at a listener position, movement of a sound source on the stage (Sect.* 2.1.4*), apparent source width (ASW) and subjective diffuseness) associated with the right hemisphere.*

2.1 Analyses of a Source Signal

2.1.1 Autocorrelation Function (ACF) of a Sound Source

The most promising signal process, in the auditory system after a rough peripheral power spectrum process, is the ACF, which is defined by

$$\Phi_{\mathrm{p}}(\tau) = \lim_{T\to\infty} \frac{1}{2T} \int_{-T}^{+T} p'(t)p'(t+\tau)\mathrm{d}t \tag{2.1}$$

Y. Ando, *Opera House Acoustics Based on Subjective Preference Theory*,
Mathematics for Industry 12, DOI 10.1007/978-4-431-55423-3_2

where $p'(t) = p(t) * s(t)$ and $s(t)$ is the sensitivity of the ear. For convenience, $s(t)$ can be chosen as the impulse response of an A-weighted network. It is worth noticing that the physical system between the sound source in front of a listener and the oval window forms almost the same characteristics as the ear's sensitivity (Ando 1985, 1998).

The normalized ACF is defined by

$$\phi_{\mathrm{p}}(\tau) = \Phi_{\mathrm{p}}(\tau)/\Phi_{\mathrm{p}}(0) \tag{2.2}$$

Thus, $\phi_{\mathrm{p}}(0) = 1$.

2.1.2 Running ACF

The short-time moving ACF or the running ACF as shown in Fig. 2.1 is calculated as (Taguti and Ando 1997).

$$\begin{aligned}\phi_{\mathrm{p}}(\tau) &= \Phi_{\mathrm{p}}(\tau;\, t,\, T)\\ &= \frac{\Phi_{\mathrm{p}}(\tau; t, T)}{[\Phi_{\mathrm{p}}(0;\, t, T)\Phi_{\mathrm{p}}(0;\ \tau + t, T)]^{1/2}}\end{aligned} \tag{2.3}$$

where

$$\Phi\mathrm{p}(\tau; t, T) = \frac{1}{2T}\int_{t-T}^{t+T} p'(s)p'(s+\tau)\mathrm{d}s \tag{2.4}$$

The normalized ACF satisfies the condition that $\phi_{\mathrm{p}}(0) = 1$.

2.1.3 Analyses of the Running ACF

In order to avoid confusion in the analyses of the running ACFs, five different signal durations analyzed are illustrated in Fig. 2.2. Resulting ACFs and the power spectra obtained by different signal durations are shown in Fig. 2.3. The direct method is obtained in the time domain. The ACF obtained by FFT also, based on the Wiener–Khintchine theorem, is acquired by a transform in the frequency domain by FFT, followed by performing an inverse FFT calculation. It is important to note that the Wiener–Khintchine theorem is mathematically satisfied only for completely periodic or infinite-length signals, but not mathematically be satisfied for a finite duration of sound signals. A variation in both ACF and power spectrum due to the different signal duration is evident (see Fig. 2.3a–t). It is not possible to find even one matched

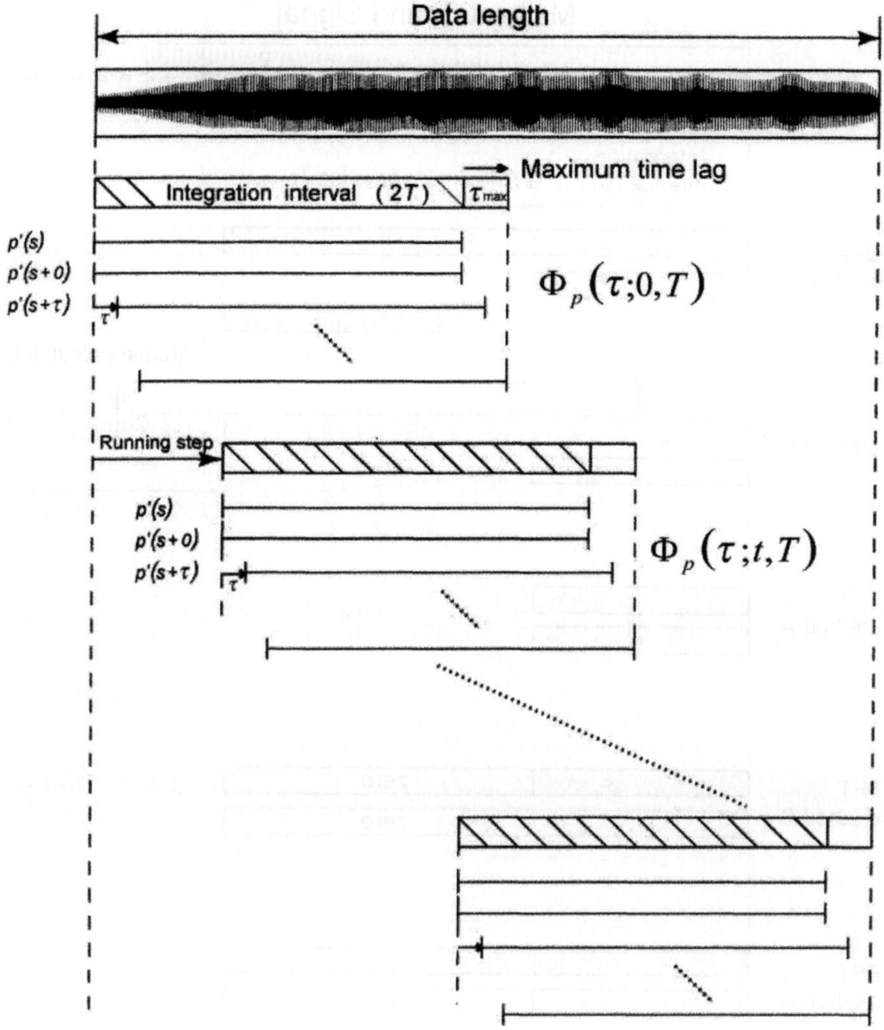

Fig. 2.1 Direct method of analyzing the running autocorrelation function (ACF) in the time domain (Kato et al. 2007)

pair of the running ACF and running power spectrum for quasi-periodic signals. Thus, we reiterate that the transform methods and their precise definitions should be carefully determined before conducting an analysis of signals.

Although "FFT method A" or "FFT method B (method to avoid circular calculation)" is usually used for the purpose of the fast computation and is accompanied by a window function such as Hamming, Hanning, or Blackman, in order to obtain the ACF corresponding to the direct method, "FFT method C" (see Fig. 2.3e) must be used. If "FFT method C" may be chosen instead of the "direct method" for

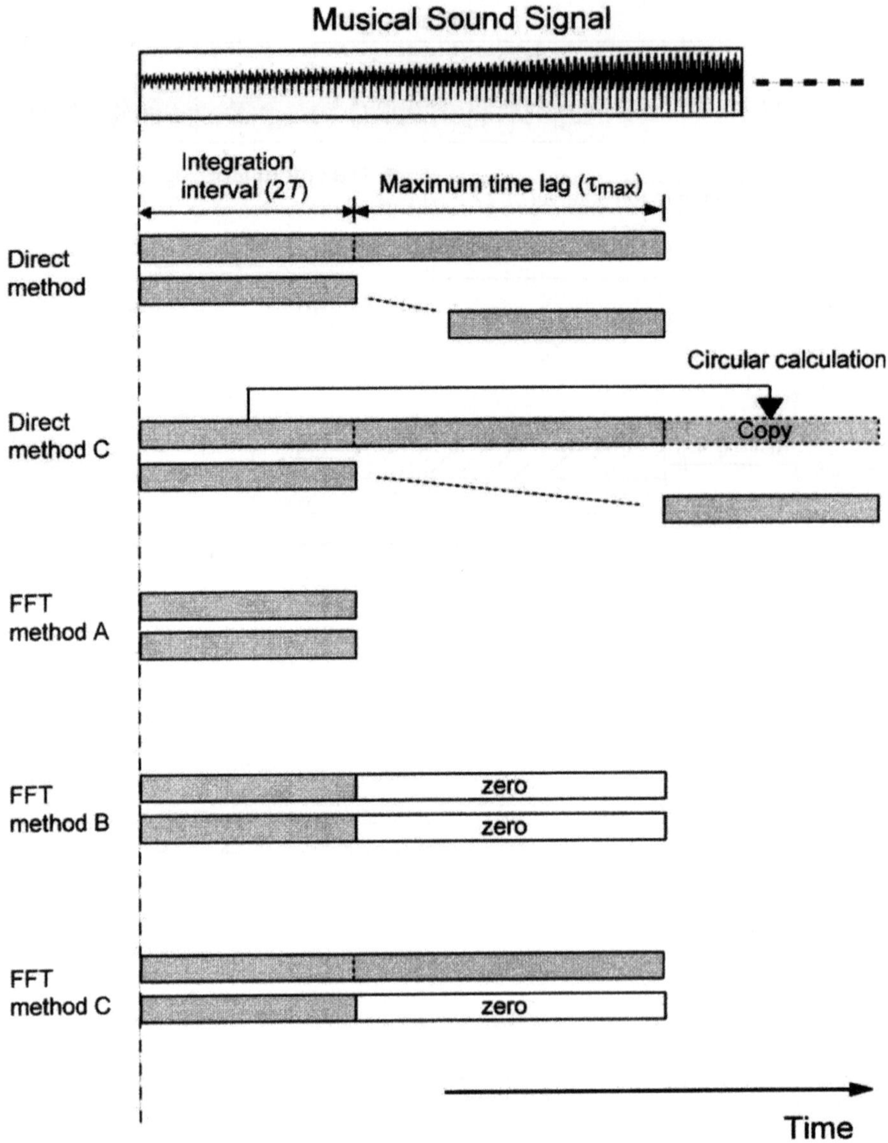

Fig. 2.2 Five different signal durations analyzed for the ACF (Kato et al. 2007)

performing a fast calculation, the segment over the maximum time lag should be deleted because of the circular calculation. Compare the result of direct method with that of FFT method C (Fig. 2.3a, e).

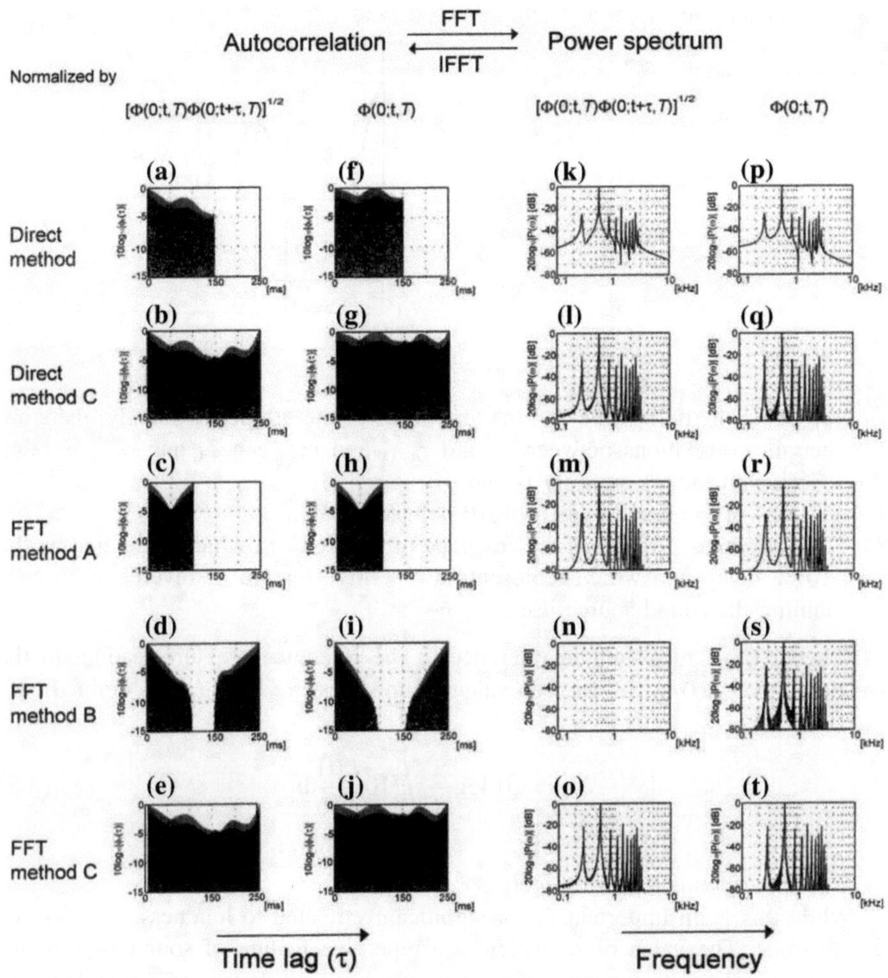

Fig. 2.3 Comparisons of the ACF and its power spectrum obtained by five different signal durations shown in Fig. 2.2. FFT methods *A* and *B* may not obtain the right ACF. FFT method *C* may be recommended for analyzing the ACF up to the maximum delay time, τ_{max} (Kato et al. 2007)

2.1.4 Temporal Factors Extracted from the Running ACF

There are significant temporal factors influencing subjective responses that can be extracted from the running ACF (Fig. 2.4):

(1) Energy represented at the origin of the delay, $\Phi_p(0)$;
(2) Fine structure, including peaks and delays. For instance, τ_1 and ϕ_1 are the delay time and the amplitude of the first peak of the ACF, respectively, τ_n and

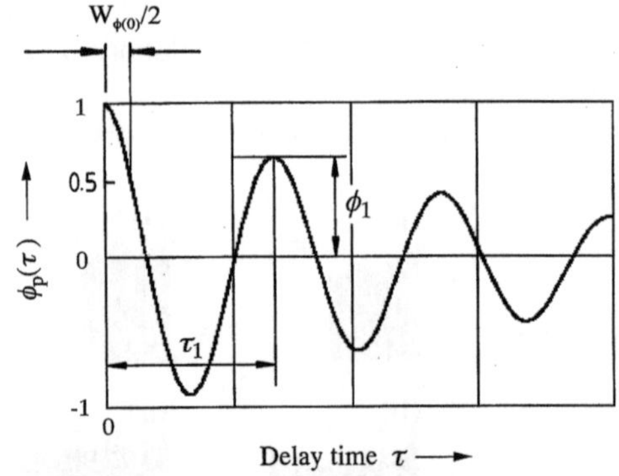

Fig. 2.4 Definition of three temporal factors extracted from the initial part of ACF

ϕ_n being the delay time and the amplitude of the nth peak. Usually, there are certain correlations between τ_1 and τ_{n+1}, and between ϕ_1 and ϕ_{n+1}, so that significant factors are only τ_1 and ϕ_1;

(3) Widths of the amplitudes of $\phi_p(0)$ defined by $W_{\phi(0)}$.
(4) The effective duration of the envelope of the ACF, τ_e, which is defined by the 10 % delay and which represents a repetitive feature or reverberation containing the sound source itself.

When $p'(t)$ is measured in reference to the reference pressure leading to the envelope level $L(t)$ in dB, the equivalent sound pressure level L_{eq}, is defined by

$$L_{eq} = 10 \log \frac{1}{T} \int_0^T 10 \frac{L(t)}{10} dt, \tag{2.5}$$

This corresponds to 10 log Φp(0).

While this is an important factor significantly related to loudness, it is not the whole story. The value of τ_e, which is a repetitive feature of sound signals, for example, is related to loudness and other subjective attributes, as is detailed later (Fig. 5.10).

In order to demonstrate a procedure for obtaining the effective duration of the analyzed short-time ACF analyzed, Fig. 2.5 shows the absolute value in the

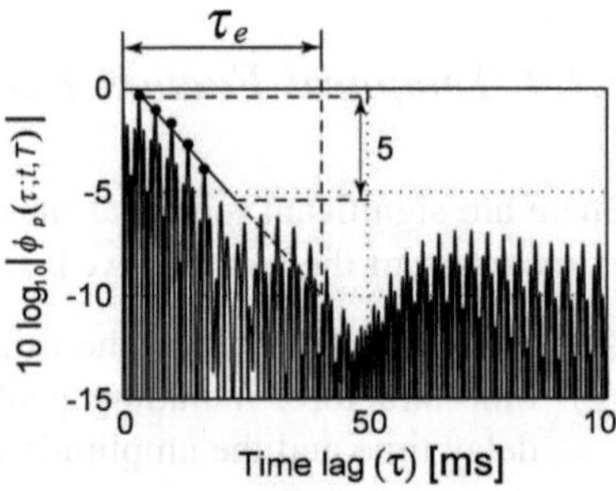

Fig. 2.5 Determination of the effective duration extracted from the running ACF

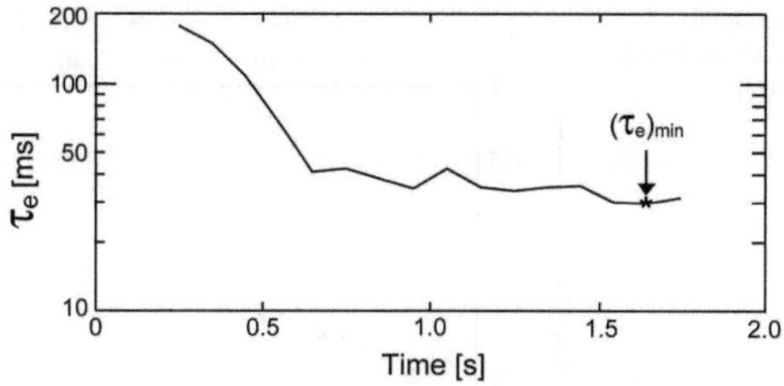

Fig. 2.6 An example of obtaining the minimum value in the effective durations extracted from the ACF

logarithmic form as a function of the delay time. The envelope decay of the initial and early part of the ACF may be fitted usually by a straight line in most cases. The effective duration of the ACF, defined by the delay τ_e at which the envelope of the ACF becomes −10 dB (or 0.1; the tenth percentile delay), can be easily obtained by the decay rate extrapolated in the range from 0 to −5 dB. When the 5 dB range is available such as for singing voice of vowels, the value of τ_e is obtained by the initial 5 ms-delay interval (Sect. 2.3).

The recommended signal duration $(2T)_r$ to be analyzed is discussed in Sect. 2.2.

2.1.5 Minimum Values of the Effective Duration Extracted from Running ACF

The minimum value of a moving τ_e, the most active part of music and speech including on and off sets of signals, containing important information and influencing subjective responses for the temporal criteria. An example of the value of $(\tau_e)_{min}$ is illustrated in Fig. 2.6.

2.2 Auditory Temporal Window

In analysis of the running ACF, so-called the "auditory-temporal window" $2T$ in Eqs. (2.3) and (2.4) must be carefully determined. The initial part of ACF within the effective duration τ_e of the ACF contains important information of the signal. In order to determine the auditory temporal window, successive loudness judgments in pursuit of the running LL have been conducted. Results are shown in Fig. 2.7 and a recommended signal duration $(2T)_r$ to be analyzed is approximately given by

$$(2T)_r \approx 30(\tau_e)_{min} \tag{2.6}$$

Fig. 2.7 Recommended signal duration to be analyzed in obtaining the ACF

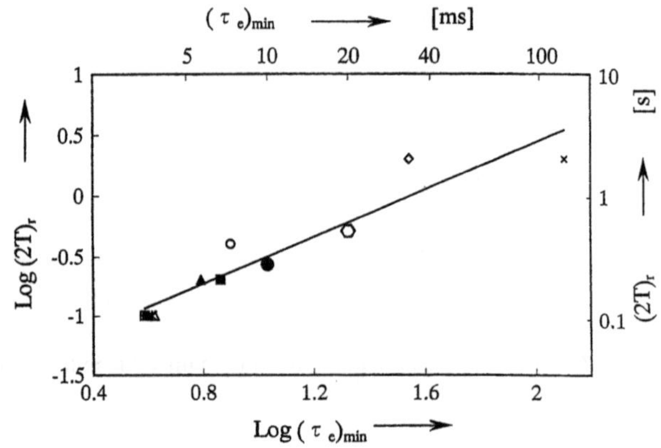

where $(\tau_e)_{min}$ is the minimum value of τ_e obtained by analyzing the ACF (Mouri et al. 2001). This signifies an adaptive temporal window "depending on the temporal activity" of the sound signal in the auditory system. For example, the temporal recommended windows differ according to music pieces $(2T)_r = 0.5–5$ s, and to the vowel $(2T)_r = 50–100$ ms and consonants $(2T)_r = 5–10$ ms in the continuous speech signal. Thus, brain might be more relaxed when listening to music than listening to speech. In other word, more concentration should be paid in listening speech than listening music.

Also, in the noise measurement, for example, the time constant represented by "fast" or "slow" of the sound level meter might be replaced by the temporal window, which is well described by the effective duration of ACF of the source signal. Note that the running step (RS), which signifies a degree of overlap of signal to be analyzed, is not critical. It may be selected as $K_2(2T)_r$, K_2 being chosen, say, in the range of 1/4–1/2.

2.3 Vocal Source Signal

In an opera house, vocal music sounds are produced on the stage. In order to demonstrate a procedure of extracting the effective duration from the running ACF analyzed, Fig. 2.5 shows the absolute value in the logarithmic form as a function of the delay time (Kato et al. 2007). The envelope decay of initial and important parts of running ACF may be fitted by a straight line in the range of 5 dB for most cases as shown in Fig. 2.8a, b. But, sometimes such a 5 dB range are not available as shown in Fig. 2.8c, so that the value of τ_e is obtained by the initial 50 ms-delay interval, as far as speech signal is concerted.

Examples of the τ_e values analyzed for vowel signals sung by a soprano are demonstrated in Fig. 2.9 with three different signal durations integrated $(2T)$.

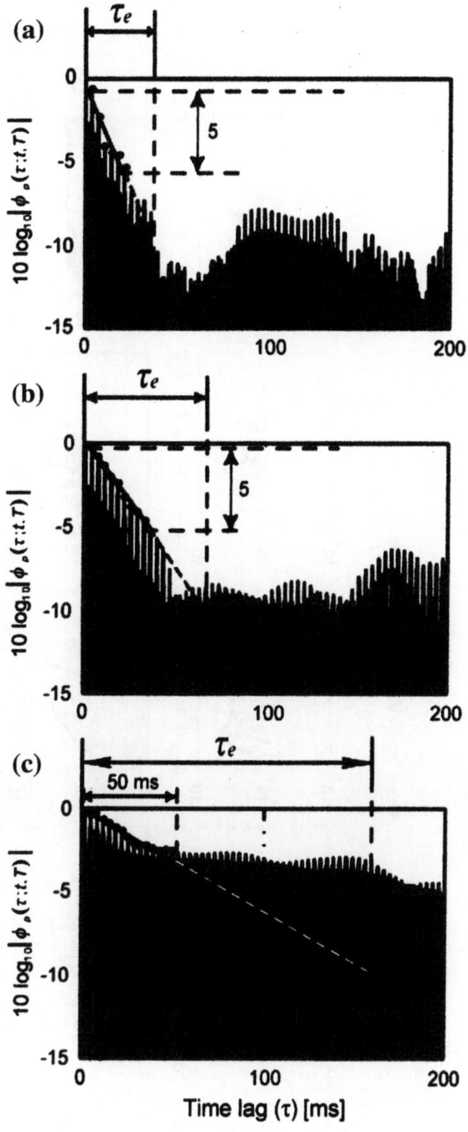

Fig. 2.8 Examples of the ACF envelope in logarithmic form and the τ_e value extracted (Kato et al. 2007)

Although, τ_e values are varied according to $2T$, however, the most important minimum value as well as local minima are independent in certain range of $2T$ for vocal signals.

Further discussion is made in Sects. 9.3 and 9.4 for blending with the sound field for listeners.

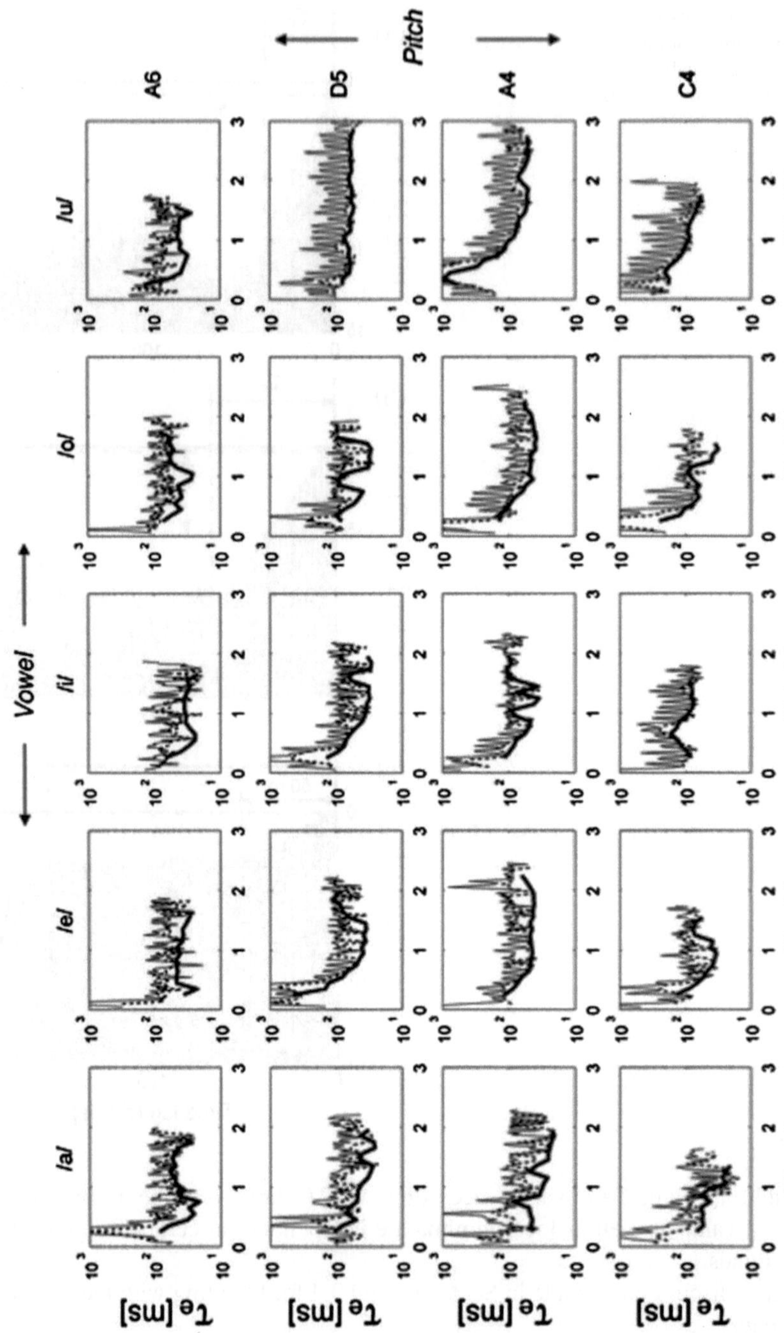

Fig. 2.9 Examples of the measured τ_e value extracted from the ACF of 20 vowels sung by a soprano singer with four different pitches obtained for three different signal durations (Kato et al. 2007). *Tine curve* $2T = 100$ ms, *dotted curve* $2T = 200$ ms, and *thick curve* $2T = 500$ ms

2.4 Running ACF of Piano Signal with Different Performance Style

We shall analyze a piano signal as a sound source in the orchestra pit. In order to examine whether or not we can control the value of τ_e *of the running ACF 2 (= 2 s) of piano signals of varying performance styles for blending with a given sound field, a piano was controlled by its performing style using a computer.*

Signals played by a piano were recorded in an anechoic chamber and analyzed (Taguti and Ando 1997). As is described above, the effective duration of running ACF, τ_e, is the fundamental time unit of the sound field (Eqs. 6.4 and 6.8; Ando 1998, 2009a). The performance style may be controlled blending the temporal factor of the sound field and the most preferred initial time delay gap between the direct sound and the first reflection, and the preferred subsequent reverberation time (Chap. 6). If the effective duration of running ACF is varied by the performing style, then musician may control it to fit the preferred temporal condition of the sound field.

Typical results of the effective duration extracted from the running ACF in changing style of piano performance—staccato and legato—are shown in Table 2.1. As is expected, staccato resulted in a short value of the effective duration, τ_e, and legato and super legato leads to long values. The use of the damper pedal creates long values of the τ_e. The minimum value of τ_e corresponds roughly to values of the note-onset duration (NOD).

Table 2.1 Various styles of piano performance and the effective duration of ACF, τ_e

Style of performance	NOD (ms)[a]	τ_e (ms)
Staccato	50	61–87
Legato	125	106–170
Super legato	160	170–233
Mixed	–	110–155

The music piece used is the opening 8 bars of exercise no. 1, Hanon Tempo mm = 120 under constant dynamics

[a] NOD is the note-on duration

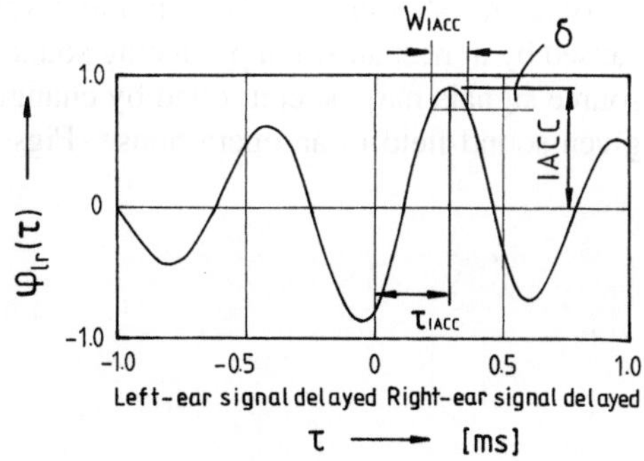

Fig. 2.10 Definitions of three spatial factors extracted from the interaural cross-correlation function (IACF)

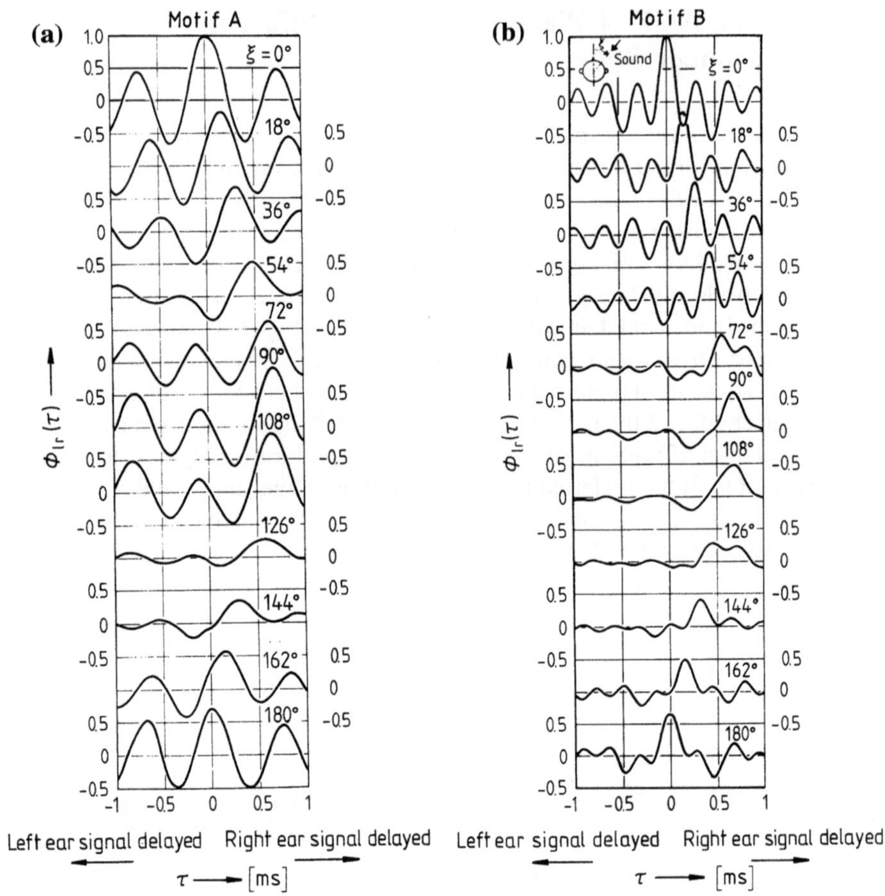

Fig. 2.11 Measured IACF in an anechoic chamber as a function of the interaural delay time and as a parameter of the horizontal angle of sound incidence (Mehrgardt and Mellert 1977). **a** Music motif A. **b** Music motif B

Staccato shortens the value of τ_e as the acuteness increases, but the value becomes no shorter than the minimum value of 60 ms. This lower limit may be caused by a mechanism in producing sound from the piano. So far, the value of τ_e of source signals may be controlled by changing the performing style blending with a given sound field in an opera house (Figs. 2.10 and 2.11).

Chapter 3
Formulation and Simulation of the Sound Field in an Enclosure

After formulating the physical system of the sound field from a source point to the two-ear entrances, a simulation system of the field for the subjective judgment incorporating temporal and spatial factors is described.

3.1 Sound Transmission from a Point Source to Ear Entrances in an Enclosure

Let us consider the sound transmission from a source point in a free field to binaural ear-canal entrances. Let $p(t)$ be the source signal as a function of time t, and $g_l(t)$ and $g_r(t)$ be impulse responses between the source point r_0 and the binaural entrances of a listener. Then the sound signals arriving at the entrances are expressed by

$$\begin{aligned} f_l(t) &= p(t) * g_l(t) \\ f_r(t) &= p(t) * g_r(t), \end{aligned} \tag{3.1}$$

where the asterisk denotes convolution. The impulse responses $g_{l,r}(t)$ consists of the direct sound and reflections $w_n(t - \Delta t_n)$ of walls in the room as well as the head-related impulse responses $h_{nl,r}(t)$, so that

$$g_{l,r}(t) = \sum_{n=0}^{\infty} A_n w_n(t - \Delta t_n) * h_{nl,r}(t), \tag{3.2}$$

where n denotes the number of reflections with horizontal angle ξ_n and elevation η_n, $n = 0$ signifies the direct sound ($\xi_0 = 0, y_0 = 0$):

$$A_0 w_0(t - \Delta t_0) = \delta(t), \quad \Delta t_0 = 0, A_0 = 1,$$

$\delta(t)$ being the Dirac delta function, A_n is the pressure amplitude of the nth reflection $n > 0$ in reference to that of the direct sound A_0; $w_n(t)$ is the impulse response of the

Y. Ando, *Opera House Acoustics Based on Subjective Preference Theory*,
Mathematics for Industry 12, DOI 10.1007/978-4-431-55423-3_3

walls for each path of reflection arriving at the listener, Δt_n is the delay time of reflection relative to that of the direct sound, and $h_{\mathrm{nl,r}}(t)$ are impulse responses for diffraction of the head and pinnae for the single sound direction of n. Therefore, Eq. (3.1) becomes

$$f_{\mathrm{l,r}}(t) = \sum_{n=0}^{\infty} p(t) * A_n w_n(t - \Delta t_n) * h_{\mathrm{nl,r}}(t) \tag{3.3}$$

When the source has a directivity characteristics, then $p(t)$ is replaced by $p_n(t)$.

3.2 Orthogonal Factors of the Sound Field

According to sound transmission from a point source to ear entrances in the sound field of an enclosure as mentioned in previous section, orthogonal factors consisting of temporal and spatial factors of the sound field may be figured out.

3.2.1 Temporal Factors of the Sound Field

The temporal factor is extracted from the set of impulse responses of the reflecting walls, $A_n(t - \Delta t_n)$ of the sound field. The amplitudes of reflection relative to that of the direct sound A_0; A_1, A_2 … are determined by the pressure decay due to the paths d_n, such that

$$A_n = d_0/d_n, \tag{3.4}$$

where d_0 is the distance between the source point and the center of the listener's head. The impulse responses of reflections to the listener are $w_n(t - \Delta t_n)$ with the delay times of Δt_1, Δt_2, … relative to that of the direct sound, which are given by

$$\Delta t_n = (d_n - d_0)/c, \tag{3.5}$$

where c is the velocity of sound (m/s).

These parameters are not physically independent, in fact the values of A_n are directly related to Δt_n in a relationship given by

$$\Delta t_n = d_0(1/A_n - 1)/c \tag{3.6}$$

In addition, the initial time delay gap between the direct sound and the first reflection Δt_1 is statistically related to Δt_2, Δt_3,…, which depend on the dimensions and the shape of the room. In fact the echo density is proportional to the square of the time delay (Kuttruff 1991). Thus, the initial time delay gap Δt_1 is regarded as a representation of both sets of Δt_n and A_n ($n = 1, 2, \ldots$).

Another parameter is the set of the impulse responses of the nth reflection, $w_n(t)$ being expressed by

$$w_n(t) = w_n(t)^{(1)} * w_n(t)^{(2)} * \cdots * w_n(t)^{(i)}, \tag{3.7}$$

where $w_n(t)^{(i)}$ is the impulse response of the ith wall existing in the path of the nth reflection from the source to the listener.

Such a set of impulse response $w_n(t)^{(i)}$ may be represented by a statistical decay rate, namely the subsequent reverberation time, T_{sub}, because $w_n(t)^{(i)}$ includes the absorption coefficient as a function of frequency. This coefficient is given by

$$\alpha_n(\omega)^{(i)} = 1 - \left|W_n(\omega)^{(i)}\right|^2 \tag{3.8}$$

It is worth noticing that as far as a single reflection is concerned, the most preferred condition of $w_n(t)^{(i)}$ is the perfectly reflection given by $\delta(t)$ (Ando 1985; see Sect. 4.2.5).

According to Sabine's formula (1900), the subsequent reverberation time is approximately calculated by

$$T_{\mathrm{sub}} \approx \frac{KV}{\bar{\alpha}S}, \tag{3.9}$$

where K is a constant (about 0.162), V is the volume of the room, S is the total surface, and $\bar{\alpha}$ is the average absorption coefficient of walls, and $\bar{\alpha}S$ is given by the summation of the absorption of each surface i, so that

$$\bar{\alpha}S = \sum_i \alpha(\omega)^{(i)} S^{(i)} \tag{3.10}$$

So far, we figured out the significant temporal factors of the sound field are:

(1) The initial delay time of the first reflection, Δt_1 given by Eq. (3.6), $n = 1$.
(2) The subsequent reverberation time, T_{sub} expressed by Eq. (3.9).

3.2.2 Spatial Factors of the Sound Field

Two sets of head-related impulse responses for two ears $h_{\mathrm{nl,r}}(t)$ constitute the spatial factors. These two response $h_{\mathrm{nl}}(t)$ and $h_{\mathrm{nr}}(t)$ play an important role in sound localization and spatial impression, but are not mutually independent objective factors. Therefore, to represent the interdependence between two impulse responses, a single factor may be introduced, i.e., the interaural cross-correlation function (IACF) between the sound signals at both ears $f_l(t)$ and $f_r(t)$, which is defined by

$$\Phi_{\mathrm{lr}}(\tau) = \lim_{T\to\infty} \frac{1}{2T} \int_{-T}^{+T} f_l'(t) f_r'(t+\tau)\mathrm{d}t, \quad |\tau| \le 1\,\mathrm{ms}, \tag{3.11}$$

where $f'_l(t)$ and $f'_r(t)$ are obtained by signals $f_{l,r}(t)$ after passing through the A-weighted network, which corresponds to the ear's sensitivity, $s(t)$. It has been shown that ear sensitivity may be characterized by the physical ear system including the external and the middle ear (Ando 1985, 1998).

The normalized interaural cross-correlation function is defined by

$$\Phi_{\mathrm{lr}}(\tau) = \frac{\Phi_{\mathrm{lr}}(\tau)}{\sqrt{\Phi_{\mathrm{ll}}(0)\Phi_{\mathrm{rr}}(0)}}, \tag{3.12}$$

where $\Phi_{\mathrm{ll}}(0)$ and $\Phi_{\mathrm{rr}}(0)$ are the ACFs at $\tau = 0$ for the left and right ear, respectively, or the sound energies arriving at both ears, and τ the interaural time delay possibly within plus and minus 1 ms. Also, from the denominator of Eq. (3.12), we obtain the binaural listening level (LL) such that,

$$\mathrm{LL} = 10 \log\left[\Phi(0)/\Phi(0)_{\mathrm{reference}}\right], \tag{3.13}$$

where $\Phi(0) = [\Phi_{\mathrm{ll}}(0)\Phi_{\mathrm{rr}}(0)]^{1/2}$ that is the geometrical mean of the sound energies arriving at the two ears and $\Phi(0)_{\mathrm{reference}}$ is the reference sound energy.

If discrete reflections arrive after the direct sound, then the normalized interaural cross-correlation is expressed by

$$\Phi_{\mathrm{lr}}^{(N)}(\tau) = \frac{\sum_{n=0}^{N} A^2 \Phi_{\mathrm{lr}}^{(n)}(\tau)}{\sqrt{\sum_{n=0}^{N} A^2 \Phi_{\mathrm{ll}}^{(n)}(0) \sum_{n=0}^{N} A^2 \Phi_{\mathrm{rr}}^{(n)}(0)}}, \tag{3.14}$$

where we put $w_n(t) = \delta(t)$ for the sake of convenience, and $\Phi_{\mathrm{lr}}^{(n)}(\tau)$ is the interaural cross-correlation of the nth reflection, $\Phi_{\mathrm{ll}}(0)^{(n)}$ and $\Phi_{\mathrm{rr}}(0)^{(n)}$ are the respective sound energies arriving at the two ears from the nth reflection. The denominator of Eq. (3.14) corresponds to the geometric mean of the sound energies arriving at the two ears.

The magnitude of the interaural cross-correlation is defined by

$$\mathrm{IACC} = |\phi_{\mathrm{lr}}(\tau)|_{\max} \tag{3.15}$$

for the possible maximum interaural time delay,

$$|\tau| \leq 1\,\text{ms}$$

For several music motifs, the long-time IACF ($2T = 35$ s) were measured for each single-reflected sound direction arriving at a dummy head (Table D.1, Ando 1985). These data may be utilized for the calculation of the IACF by Eq. (3.14). For example, measured values of the IACF using music motifs A and B are shown in Fig. 2.11a, b.

The interaural delay time, at which the IACC is defined as shown in Fig. 2.10, is the τ_{IACC}. Thus, both the IACC and τ_{IACC} may be obtained at the maximum value of IACF. For a single source signal arriving from the horizontal angle ξ defined by τ_ξ, the interaural time delay corresponds to τ_{IACC}. When it is observed $\tau_{\text{IACC}} = 0$ in an opera house, then usually a frontal sound image and a well left- and right-balanced sound field are perceived (the preferred condition).

The width of the IACF, defined by the interval of delay time at a value of δ below the IACC, corresponding to the JND of the IACC, is given by the W_{IACC} (Fig. 2.10). A well-defined directional impression corresponding to the interaural time delay τ_{IACC} is perceived when listening to a white noise with a sharp peak in the IACF with a small value of W_{IACC}. Thus, the apparent source width (ASW) may be perceived as a directional range corresponding mainly to the W_{IACC}. On the other hand, when listening to a sound field with a low value for the IACC < 0.15, then a subjectively diffuse sound is perceived (Damaske and Ando 1972). These four factors, LL, IACC, τ_{IACC} and W_{IACC} are independently related to spatial percepts such as subjective diffuseness and the ASW (Sect. 5.2). These four factors, LL, IACC, τ_{IACC} and W_{IACC} are independently related to spatial sensations such as subjective diffuseness and the ASW (Sect. 6.2; Ando et al. 1999; Ando 2002).

Significant spatial factors of the sound field for subjective preference are extracted from the IACF.

(1) The binaural listening level (LL) is obtained accurately as defined by Eq. (3.13).
(2) The IACC is defined by Eq. (3.15) as defined in Fig. 2.10.
(3) The interaural delay time is the τ_{IACC} at which the IACC.

3.2.3 Auditory Time Window for the IACF Processing

These spatial sensations may be judged immediately when we come into a sound field, because our binaural system may process for the IACF in the short-time window as discussed below. This is quite difference from the adaptive temporal window for the sound signals, which varies due to the effective duration of the ACF of the sound source signal.

When a sound signal is moving in the horizontal direction on the stage, we must select a suitable short "time window" $2T$ in analyzing the running IACF which depend on the speed of moving image of the sound localization. The range of τ_{IACC} extracted from the IACF can describe the range of such a moving image. It is obvious that the range of τ_{IACC} cannot be obtained, when the integration interval ($2T$) of the IACF is longer than the period of movement; on the other hand, the value of τ_{IACC} is too fluctuated to be determined, when $2T$ is selected shorter than the possible maximum value of $\tau_{IACC} < 1$ ms. For a sound source moving sinusoidally in the horizontal plane with less than 0.2 Hz, $2T$ may be selected in a wide range from 30 to 1,000 ms. And, when a sound source is moving in a range of 4.0 Hz, $2T = 30$–100 ms is acceptable (Mouri et al. unpublished). To obtain reliable results, it is recommended for such a temporal window for the IACF covering a wide range of movement velocity in the horizontal localization, to be fixed, say, about 30 ms.

For the sound source fixed on the stage in an opera house, for example, the value of ($2T$) may be selected longer than 1.0 s for the measurement of spatial factors at each audience seat.

3.3 Simulation of Sound Localization

The directional information in simulating the sound field in an opera house can be realized by means of spatial factors that are extracted from the IACF. Schroeder (1962) first simulated sound localization in the horizontal plane by use of two loudspeaker reproduction system. To make the perception correspond precisely to the actual direction of a sound source located at any position in a three-dimensional space, a general system considering asymmetry of our head and pinnae (Ando et al. 1973) is described as follows.

Referring to the lower part of Fig. 3.1, let the pressure impulse response for the paths from the two loudspeakers L_1 and L_2 to the entrances of the left and right ear canals be $h_{l,r1}(t)$ and $h_{l,r2}(t)$, respectively. Then, the pressures to be reproduced at the two entrances are expressed by

$$\begin{aligned} f_l(t) &= x_1(t) * h_{l1}(t) + x_2(t) * h_{l2}(t) \\ f_r(t) &= x_1(t) * h_{r1}(t) + x_2(t) * h_{r2}(t), \end{aligned} \tag{3.16}$$

where x_1 and x_2 are the input signals supplied for the loudspeakers L_1 and L_2, respectively.

Fourier transforming both sides of Eq. (3.16) yields

$$\begin{aligned} F_l(\omega) &= X_1(\omega)H_{l1}(\omega) + X_2(\omega)H_{l2}(\omega) \\ F_r(\omega) &= X_1(\omega)H_{r1}(\omega) + X_2(\omega)H_{r2}(\omega) \end{aligned} \tag{3.17}$$

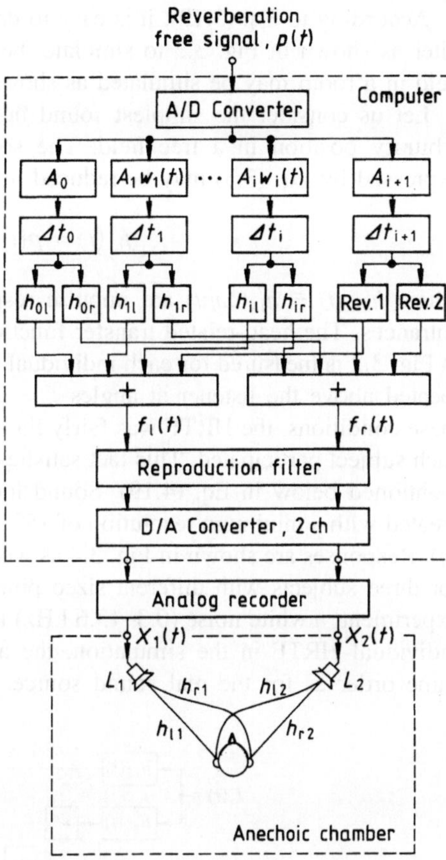

Fig. 3.1 A system of simulating the sound field with the direct sound and two early reflections and two incoherent reverberators in an enclosure and a reproduction system with two loudspeakers

Thus, the input signals to the loudspeakers are given by

$$\begin{aligned} X_1(\omega) &= [F_1(\omega)H_{r2}(\omega) - F_r(\omega)H_{l2}(\omega)]D(\omega)^{-1} \\ X_2(\omega) &= [F_r(\omega)H_{l1}(\omega) - F_l(\omega)H_{r1}(\omega)]D(\omega)^{-1}, \end{aligned} \tag{3.18}$$

where $D(\omega) = H_{l1}(\omega)H_{r2}(\omega) - H_{r1}(\omega)H_{l2}(\omega)$.

Therefore, the signals in the time domain to be fed into two loudspeakers are obtained by the inverse Fourier transform, such that

$$\begin{aligned} x_1(t) &= [f_1(t) * h_{r2}(t) - f_r(t) * h_{l2}(t)] * \mathrm{d}(t) \\ x_2(t) &= [f_2(t) * h_{l1}(t) - f_1(t) * h_{r1}(t)] * \mathrm{d}(t), \end{aligned} \tag{3.19}$$

where d(t) is the inverse Fourier transform of $D(w)^{-1}$. The necessary and sufficient condition for a unique solution is $D(\omega) \neq 0$, throughout the frequency range reproduced.

According to Eq. (3.19), it is easy to draw a block diagram of the reproduction filter as shown in Fig. 3.2 to simulate the sound localization, and thus the sound field in a room may be simulated as shown in Fig. 3.1.

Let us consider the simplest sound field, a single sound source located at an arbitrary position in a free field. The sound pressure at the two ear entrances expressed by Eq. (3.3) may be reduced into a simple form, such that

$$f_{\mathrm{l,r}}(t) = p(t) * h_{\mathrm{nl,r}}(t), \tag{3.20}$$

where $h_{\mathrm{nl,r}}(t) \equiv h_{\mathrm{l,r}}(\xi,\eta;t)$ are impulse responses between the source and the ear entrances. The head-related transfer function (HRTF) required for the filter shown in Fig. 3.2 is measured for each individual. In the experiment, two loudspeakers are located above the listener at angles $\xi = \pm 30°$, $\eta = 90°$ as shown in Fig. 3.3. In these conditions, the HRTF was fairly flat with no zeros and no significant dips for each subject participated. This fact satisfies the condition for the unique solution as mentioned below in Eq. (4.19). Sound localization with external sound image is created with a minimum resolution of 15°, in horizontal plane (ξ) and median plane (η). Responses are shown in Fig. 3.4 as well as localization with real sound sources for three subjects with different sized pinnae (Morimoto and Ando 1980). In this experiment, a white noise (0.3–13.6 kHz) is presented as a source signal. By use of individual HRTF in the simulation, the accuracy of localization was almost the same order as for the real sound source. If we apply the HRTF from the other

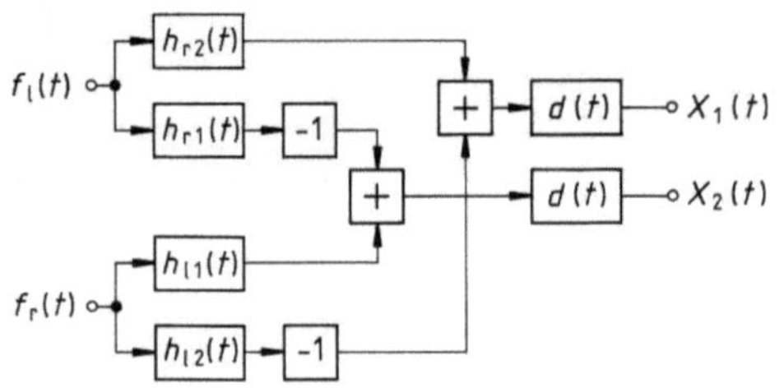

Fig. 3.2 Reproduction filter for two loudspeaker system for the two ears

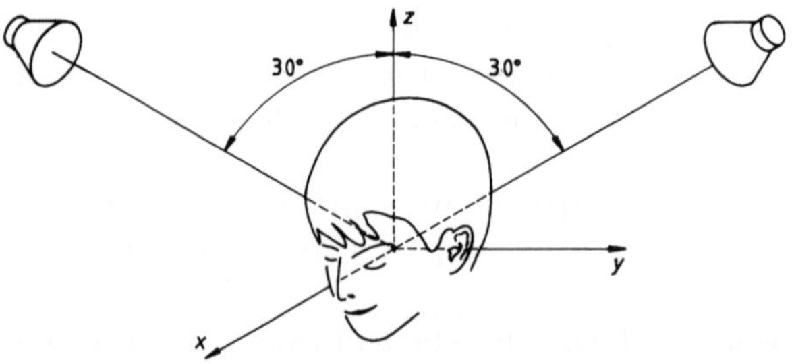

Fig. 3.3 Location of two loudspeakers for simulating sound localization in three-dimensional space

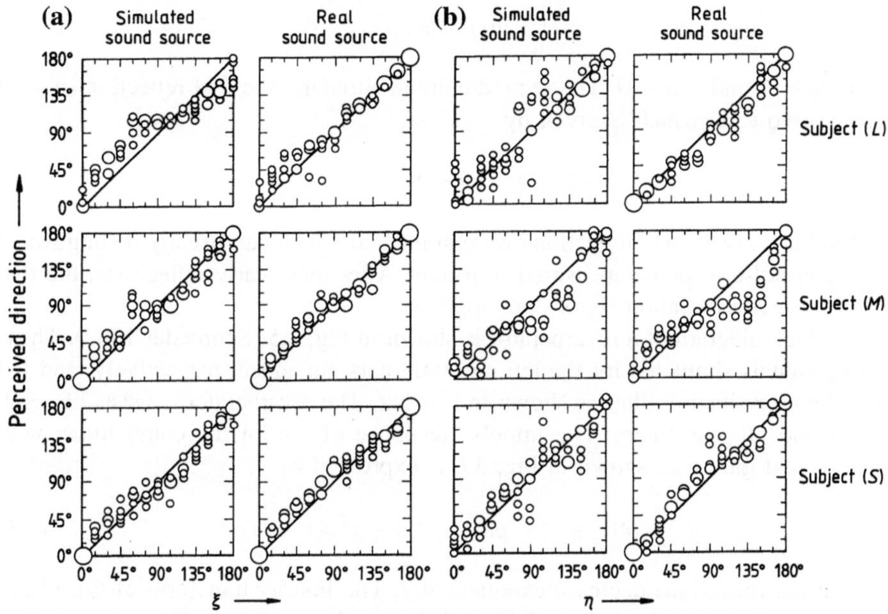

Fig. 3.4 Results of sound localization tests by three listeners with different sized pinnae for simulated sound source and real sound source (Morimoto and Ando 1980). **a** Horizontal plane. **b** Median plane

person, then the subject's accuracy in localization is generally decreased and, in some cases, localization is not possible.

Applying the reproduction system, the sound field in a real opera house with scattered and diffusing elements may be evaluated subjectively. When two loudspeakers are located at closely spaced, for example, $\xi = \pm 5°$, $\eta = 0°$, that is known as a "stereo-dipole" system (Kirkeby et al. 1998). A merit of the nonindividualized system is uncritical in localization even though head movement during listening. Also, it is reported by use of this system that one can even distinguish an auditory distance from judgment of a room size perception (Martignon et al. 2005).

3.4 Simulation of the Reverberant Sound Field

An example of a simulation system for the sound field in an opera house is shown in Fig. 3.1, based on Eq. (3.3). A reverberation free vocal or orchestra music signal is applied for $p(t)$. The program provides the amplitude and delay time of early reflections including directional information, and the subsequent reverberation. All calculated relative to the direct sound ($n = 0$). As shown in the first column of the upper half part of Fig. 3.1, the direct sound is simulated by using only the HRTF to the two ears for the frontal direction, i.e.,

$$p(t) * h_{01,\mathrm{r}}(t) \tag{3.21}$$

with $A_0 = 1$ and $\Delta t_0 = 0$. The second column simulates the first reflection ($n = 1$) for the two ears, which is given by

$$p(t) * A_1 w_1(t - \Delta t_1) * h_{1,1,\mathrm{r}}(t) \tag{3.22}$$

Similarly, two early reflections were simulated which can usually distinguished in the impulse response measured in rooms. After these early reflections, the two incoherent reverberation signals are applied.

A block diagram of a reverberator is shown in Fig. 3.5 (Schroeder 1962). These sound signals simulated for the left and right ears are added, respectively, and fed into the reproduction filter as shown in Fig. 3.2. The reverberator consists of comb filters and all-pass filters. The impulse response of one of the comb filters with delay τ and gain g as shown in Fig. 3.6 is expressed by

$$h(t) = \delta(t - \tau) + g\delta(t - 2\tau) + g^2\delta(t - 3\tau) + \cdots \tag{3.23}$$

so that the reflections decrease exponentially. The Fourier transform of Eq. (3.23) gives corresponding frequency characteristics, such as

$$H(\omega) = e^{-j\omega\tau} + ge^{-j2\omega\tau} + g^2 e^{-j3\omega\tau} + \cdots = e^{-j\omega\tau}/(1 - ge^{-j\omega\tau}) \tag{3.24}$$

The absolute vale of $H(\omega)$, which is given by

$$|H(\omega)| = 1/\left[1 + g^2 - 2g\cos\omega\tau\right]^{1/2} \tag{3.25}$$

is shown in the lower part of Fig. 3.6. The amplitude as a function of frequency presents a comb with periodic structure. For $\omega = 2n\pi/\tau$, $n = 0, 1, 2\ldots$ and $g > 0$, it has the maxima, so that

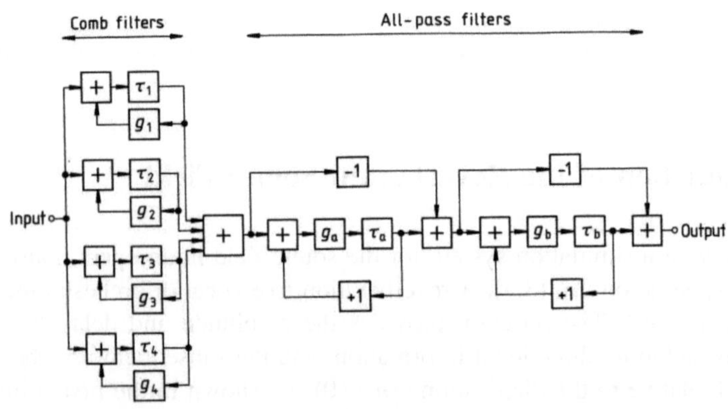

Fig. 3.5 Reverberator with four comb filters controlling the subsequent reverberation time and two all-pass filters simulating the density of reflections (Schroeder 1962)

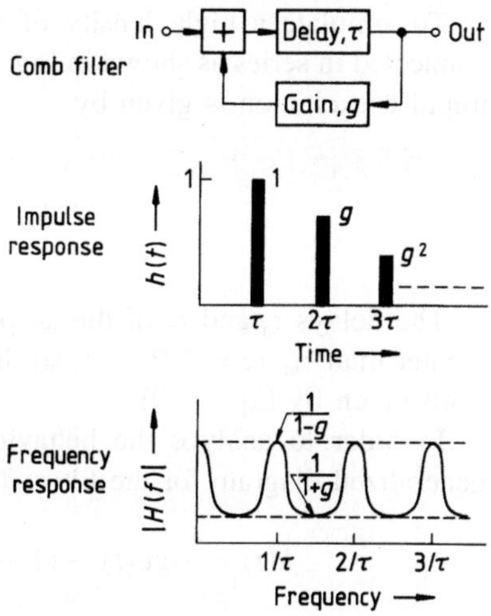

Fig. 3.6 Comb filter, the impulse response and the frequency characteristics (Schroeder 1962)

$$|H(\omega)|_{\min} = 1/(1-g) \tag{3.26}$$

And for $\omega = (2n+1)\pi/\tau$, $n = 0, 1, 2\ldots$ the minima

$$|H(\omega)|_{\max} = 1/(1+g) \tag{3.27}$$

Thus, the ratio between the maxima and minima yields

$$|H(\omega)|_{\max}/|H(\omega)|_{\min} = (1+g)/(1-g) \tag{3.28}$$

For example, if $g = 0.85$, then the ratio is 12.3 or 22 dB. This produces a "colored" and "fluttered" quality. By use of several different comb filters connected in parallel as shown in Fig. 3.5, highly irregular frequency response may avoid such an undesired phenomenon. The reverberation time is given by the loop gains $g_1, g_2, \ldots g_M$ and delays $\tau_1, \tau_2 \ldots \tau_M$ of the different comb filters. A sound level decay by -20 log (g_m) (dB) for every trip around the feedback loop τ_m gives

$$T_m = 60\tau_m/[-20 \log(g_m)] = 3\tau_m/[-\log(g_m)], \quad m = 1, 2, \ldots, M \tag{3.29}$$

And, the subsequent reverberation time yields

$$T_{\text{sub}} = [T_m]_{\max} \tag{3.30}$$

Note that the reverberation time as a function of frequency can be realized by the impulse response $g_m(t)$ or its Fourier transform $G_m(\omega)$, which corresponds to the transfer function for reflection from the boundary wall.

To simulate a high density of reflections of order t^2, two all-pass filters are connected in series as shown in Fig. 3.5. The density of reflections at time t after the impulse excitation is given by

$$n_e(t) = \frac{1}{2}\sum_{m=1}^{M}\frac{1}{\tau_m}\frac{1}{\tau_a\tau_b}t^2 \qquad (3.31)$$

The delays τ_a and τ_b of the all-pass filters should be chosen as τ_a and τ_b much greater than $\tau_m, m = 1, 2...M$, so that they do not influence the reverberation time itself given by Eq. (3.29).

In order to analyze the behavior of the all-pass filter, Fig. 3.7 represents a generalized diagram for the filter. The impulse response is given by

$$h(t) = -g\delta(t) + (1 - g^2)[\delta(t - \tau) + g\delta(t - 2\tau) + \cdots] \qquad (3.32)$$

Taking Eq. (3.24) into account, the Fourier transform of Eq. (3.32) yields

$$H(\omega) = -g + (1 - g)e^{-j\omega\tau}/\left(1 - ge^{-j\omega\tau}\right) = e^{-j\omega\tau}(1 - ge^{j\omega\tau})/\left(1 - ge^{-j\omega\tau}\right)$$

Thus,

$$|H(\omega)| = 1.0$$

for all frequencies.

In the reverberator shown in Fig. 3.5, if we set $g_a = g_b = 1/\sqrt{2}\,(\approx 0.7)$, then the all-pass filter is realized. It is worth noticing that preferred spectra of the single reflection and of reverberation time are just "flat" (Ando, Sect. 4.2.5 in 1985, Sect. 6.2.2 in 1998).

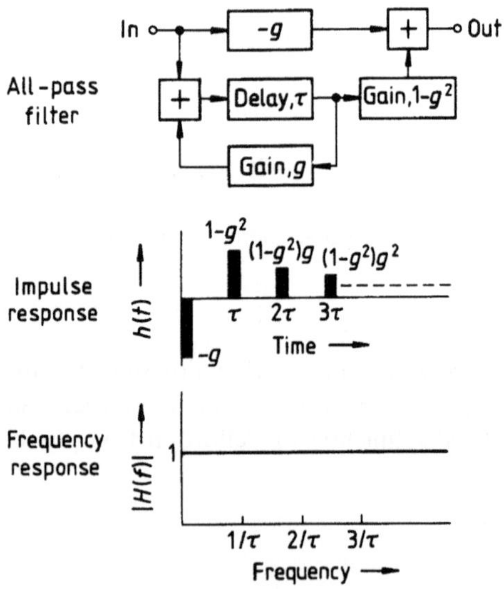

Fig. 3.7 All-pass filter, its impulse response and spectrum (Schroeder 1962)

Chapter 4
Model of Auditory-Brain System

Based on neural activities in the auditory-brain system of human, a signal processing model including the specialization of human cerebral hemispheres is described. The temporal factors associated with the left cerebral hemisphere may be extracted from the ACF processors. And the spatial factors associated with the right hemisphere may be extracted from the IACF processor for the signal arriving at the two ear entrances. Therefore, temporal and spatial primary percepts have been well described by applying these temporal and spatial factors, respectively (Ando 2002, 2003, 2009a). Subjective preference is the overall response judging incorporation with activities of the right and left hemispheres.

4.1 Neural Evidences in Auditory-Pathway and Brain System

4.1.1 Physical System

The human ear sensitivity to the sound source in front of the listener has been essentially formed by the physical system of sound and vibration from the source point to oval window of cochlea (Ando 1985, 1998). The transfer function of such cascade systems includes the human head and pinna, the external canal and the eardrum, and born chain. In order to determine temporal and spatial factors for the sake of convenience, the A-weighting network may be applied, which corresponds to the ear sensitivity.

4.1.2 ABR from the Left and Right Auditory Pathways

The records of auditory brainstem responses (ABR) imply the following:

① Amplitudes of wave $I_{l,r}$ and $III_{l,r}$ correspond roughly to the sound pressure level as a function of the horizontal angle of incidence to listener (ξ) (Ando et al. 1991).

Y. Ando, *Opera House Acoustics Based on Subjective Preference Theory*, Mathematics for Industry 12, DOI 10.1007/978-4-431-55423-3_4

② Amplitudes of waves $II_{l,r}$ and $IV_{l,r}$ correspond roughly to the sound pressure level as a function of the contra horizontal angle ($-\xi$). Thus, the sound pressure level well corresponds to the ABR amplitude, and the signal in the auditory pathways interchanges thrice before being fed into the brain (Fig. 4.1).

③ Figure 4.2 shows values of the magnitude of IACF, and the ACF at the time origin. These were measured at the two ear entrances of a dummy head as a function of the horizontal angle after passing through the A-weighting networks. Figure 4.3 shows results of analyses of ABR indicating possible neural activities around the inferior colliculus, which correspond well to the values of IACC.

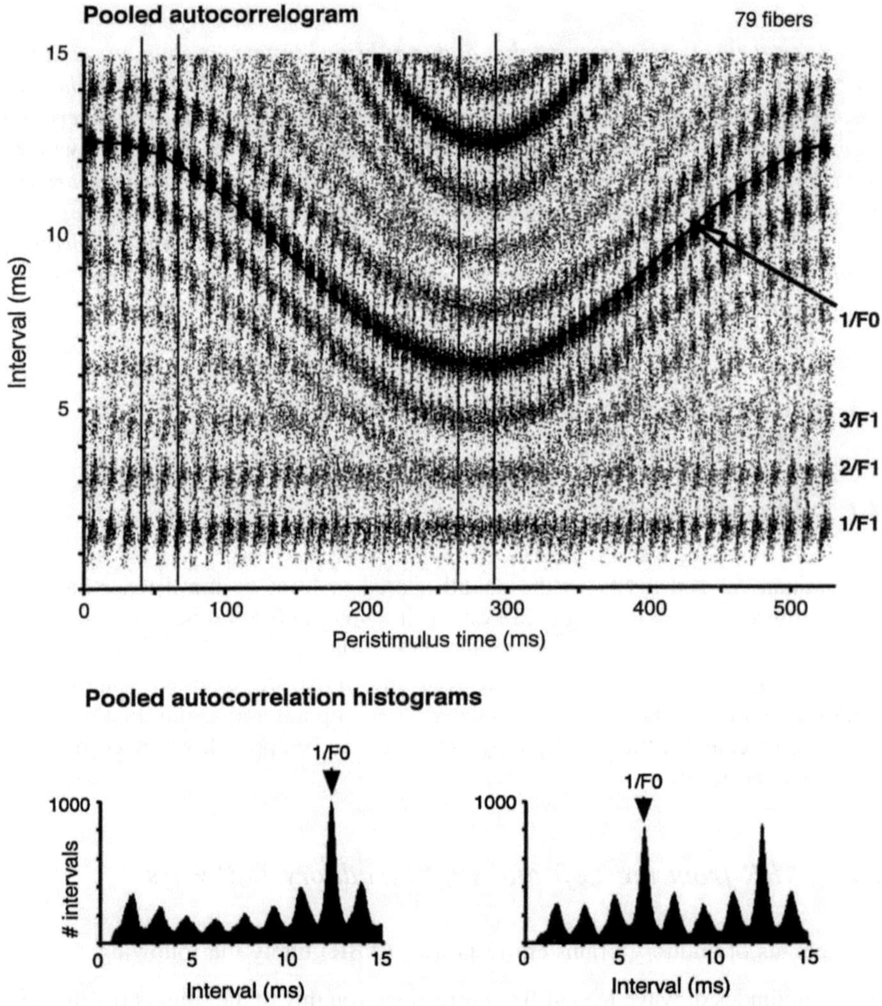

Fig. 4.1 Autocorrelograms and autocorrelation histograms in response to a variable F_0 from 80 to 160 Hz of a single-formant vowel (Cariani and Delgutte 1996a). *Upper* Pooled Autocorrelograms for the vowel. *Lower* Smoothed Autocorrelograms at two different regions

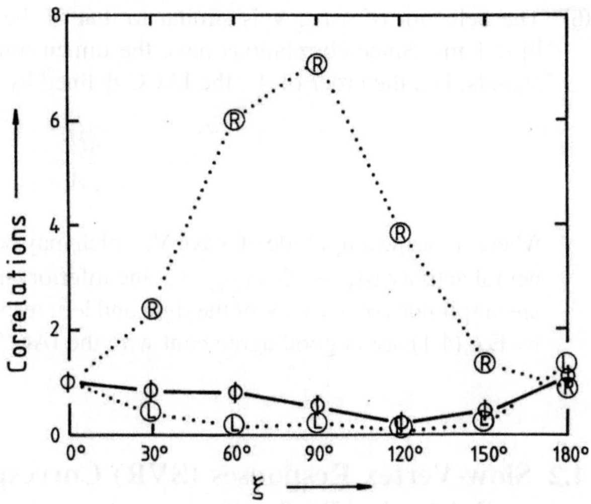

Fig. 4.2 Measured correlations of sound signal arriving at the left and right ear entrances of a dummy head. Ⓛ $\Phi_{ll}(0)$, Ⓡ $\Phi_{rr}(0)$ and $\Phi : |\Phi_{lr}(\tau)|_{max}, |\tau| \leq 1$ ms

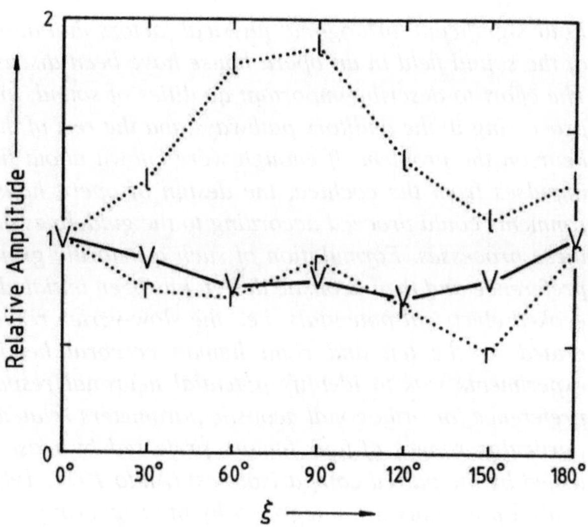

Fig. 4.3 Averaged amplitudes of the auditory brainstem response (ABR), the wave IV_l (symbol l), and the wave IV_r (symbol r) normalized to the amplitude at the frontal incidence. Symbol "V" indicates the averaged amplitudes of waves V_l and V_r. Amplides are normalized to their values at the frontal incidence (four subjects)

④ The averaged amplitudes of waves IV (left and right) and averaged amplitudes of wave V that were both normalized to the amplitudes at the frontal incidence ($\xi = 0°$) are shown in Fig. 4.3.
Although we cannot make a direct comparison between the results in Figs. 4.2 and 4.3, we can point out that the relative behavior of wave IV(l) in Fig. 4.3 is similar to Ⓡ or $\Phi_{rr}(0)$ in Fig. 4.2, which was measured at the right-ear entrance R.

⑤ Also, the relative behavior of wave IV(r) is similar to Ⓛ or $\Phi_{ll}(0)$ at the left-ear entrance L. In fact, amplitudes of wave IV (left and right) are proportional to $\Phi_{rr}(0)$ and $\Phi_{ll}(0)$, respectively, due to the interchange of signal flow.

⑥ The behavior of wave V is similar to that of the maximum value, $|\Phi_{lr}(\tau)|_{max}$, $|\tau| < 1$ ms. Since correlations have the dimensions of the power of the sound signals, i.e., the order of A^2, the IACC defined by Fig. 2.10 may correspond to

$$P = \frac{A_V^2}{[A_{IV,r}\,A_{IV,l}]} \tag{4.1}$$

where A_V is the amplitude of wave V, which may be reflected by the "maximum" neural activity ($A_V^2 \approx |\Phi_{lr}(\tau)|_{max}$) at the inferior colliculus. Also, $A_{IV,r}$ and $A_{IV,l}$ are amplitudes of wave IV of the right and left, respectively. The results obtained by Eq. (4.1) are in good agreement with the IACC (Ando 1998).

4.2 Slow-Vertex Responses (SVR) Corresponding to Subjective Preference

Four significant, orthogonal physical factors that describe time and space criteria of the sound field in an opera house have been discussed in the previous chapter. The effort to describe important qualities of sound, in terms of neural information processing in the auditory pathways and the rest of the brain, has been brought to bear on the problem. If enough were known about how the brain analyzes nerve impulses from the cochlea, the design of opera houses and other acoustic environments could proceed according to the guidelines derived from the knowledge of these processes. Formulation of such a neurally grounded strategy for subjective preference and then acoustic design has been initiated through a study of auditory-evoked electrical potentials, i.e., the slow-vertex responses (SVR), which are generated by the left and right human cerebral hemispheres. The goal of these experiments was to identify potential neuronal response correlates of subjective preference for orthogonal acoustic parameters related to sound fields. We had that particular ranges of four factors preferred by most listeners, which were established by the paired-comparison test (Ando 1977, 1983, 1985, 1998) and auditory evoked potentials are integrated by the triggering technique, so that reliable data may result. Here, we integrated the SVR for paired stimuli in a similar manner obtaining the scale value of subjective preference based on the paired-comparison method. The SVR is the response of brain after the ABR and is assumed to be an extracted factor correlating to the subjective preference. It has been found that neuronal responses correlate to the subjective preference in the following sections. Temporal factors of the sound field reflect reverb such as the initial time delay gap between the direct sound and the first reflection (Δt_1) and subsequent reverberation time (T_{sub}) are deeply associated with the left hemisphere. The typical spatial factor of the sound field, IACC, and the binaural listening level (LL) is associated with the right hemisphere (Table 4.1*; Ando 2003).*

Table 4.1 Hemispheric specializations determined by AEP, EEG, and MEG of the left and right hemispheres for temporal and spatial factors of the sound field, respectively

Factors changed	AEP (SVR) $A(P_1 - N_1)$	EEG, ratio of ACF τ_e values of α-wave	AEP (MEG) N1m	MEG, ACF τ_e value of α-wave
Temporal				
Δt_1	L > R (speech)[a]	L > R (music)		L > R (speech)
T_{sub}	–	L > R (music)	–	
Spatial				
LL	R > L (speech)	–	–	
IACC	R > L (vowel/*a*/)	R > L (music)[b]	R > L (band noise)[c]	
	R > L (band noise)			
τ_{IACC}			R > L (band noise)[c]	
Head-related transfer functions			R > L (vowels)[d]	

[a] Sound source used in experiments is indicated in the bracket
[b] The flow of EEG α-wave from the right hemisphere to the left hemisphere for music stimulus in change of the IACC was determined by the CCF $|\phi(\tau)|_{max}$ between α-waves recorded at different electrodes
[c] Soeta and Nakagawa (2006)
[d] Palomaki et al. (2002)

We found neural response correlates of subjective preference in the latency of SVR waves. The top plots of Fig. 4.4 summarize the relationship between subjective preference scale values and three acoustic parameters (LL, Δt_1, and IACC). Applying the paired method of stimuli, both SVR and the subjective preference for sounds fields were investigated as functions of SL and Δt_1. The source signal was the 0.9 s speech segment. The lower part of this figure indicates the appearance of latency components.

① As shown in the left and center columns in this figure, the neural information related to subjective preference appeared typically in an N_2-latency of 250–300 ms, when SL and Δt_1 were changed.

② Further details of the latencies for both the test sound field and the reference sound field, when Δt_1 was changed, are shown in Fig. 4.5. The parallel latencies at P_2, N_2, and P_3, were clearly observed as functions of the delay time Δt_1. However, latencies for the reference sound field ($\Delta t_1 = 0$) in the paired stimuli are found to be relatively shorter, while the latencies for the test sound field with $\Delta t_1 = 25$ ms, the most preferred delay, become the longest. This may indicate a kind of relative behavior of the brain, underestimating the reference sound field when the test sound field in the pair is the most preferred condition.

③ Relatively long-latency responses are always observed in the subjectively preferred range of each factor.

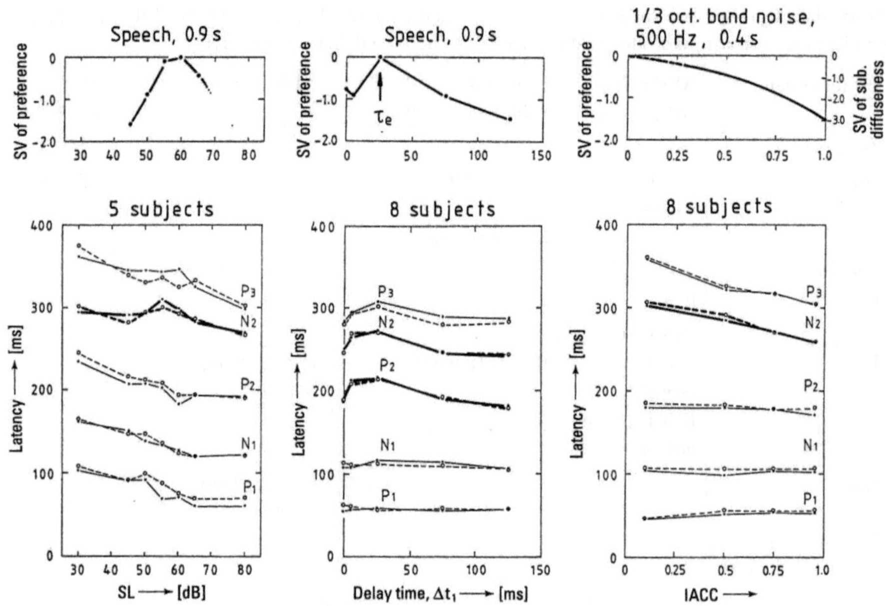

Fig. 4.4 Relationships between averaged latencies of SVR and scale values of subjective preference for three factors of the sound field. *Line* left hemisphere; *hyphenated line* right hemisphere. **a** As a function of the sensation level (SL). **b** As a function of the delay time of reflection, Δt_1. **c** As a function of the IACC

④ Thus, the difference in N_2-latencies over both hemispheres in response to a pair of sound fields contains almost the same information obtained from PCT as for the subjective preference. In general, the subjective preference may be judged in the direction of maintaining life; therefore, it may appear in neuronal response as a primitive response.
The right column of Fig. 4.4 shows the effects of varying the IACC using the 1/3-octave-band noise (500 Hz) (Ando et al. 1987; Ando 1992). On the upper part, the scale value of the subjective diffuseness is indicated as a function of IACC. The scale value of the subjective preference has a similar behavior plotted against the IACC, when speech or music signals are presented as described in Sect. 3.2.

⑤ The information related to subjective diffuseness or subjective preference, therefore, appears in the N_2-latency, ranging from 260 to 310 ms, in which a tendency for an increasing latency while decreasing the IACC was observed for eight subjects (except for the left hemisphere of one subject). As already indicated in Fig. 4.4, the relationship between the IACC and the N_2-latency was found to be linear and the correlation coefficient between them was −0.99 ($p < 0.01$).

⑥ Furthermore, let us look at the behavior of early latencies of P_1 and N_1. These were almost constant when the delay time and the IACC were changed.

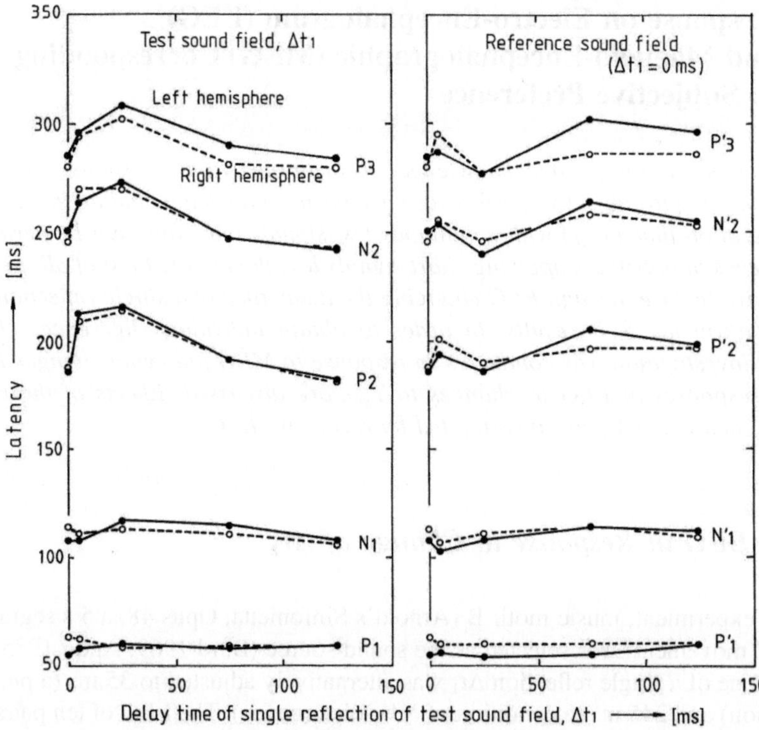

Fig. 4.5 Averaged latencies for the test sound field and the reference sound field for paired stimuli, as a function of the delay time of the reflection, Δt_1. *Line* Left hemisphere; *hyphenated line* right hemisphere. Maximum latencies of P_2, N_2 and P_3 are found at $\Delta t_1 = 25$ ms for the test sound field, while relatively short latencies of P'_2, N'_2 and P'_3 are observed for the reference sound field. This is a typical brain activity showing "relativity"

However, the information related to SL or loudness may be found typically at the N_1-latency. This tendency agrees well with the results of Botte et al. (1975).

⑦ Consequently, from 40 to 170 ms of the SVR, the hemispheric dominance may be found for the amplitude component, which may be related to respective functional specializations of the hemispheres. Early latency differences corresponding to the SL may be found in the range of 120–170 ms.

⑧ Finally, we found that the N_2-latency components in the delay range between 200 and 310 ms may correspond well with the subjective preference relative to the listening level, the time delay of the reflection, and indirectly the IACC.

⑨ Since the longest latency was always observed for the most preferred condition, one might speculate that the brain is most relaxed at the preferred condition, and that this causes the observed latency behavior to occur. Such a longest period may relate to the alpha wave having the longest period in electroencephalography (EEG) and magnetoencephalography (MEG) during the human waking stage as discussed below.

4.3 Response on Electro-Encephalogram (EEG) and Magneto-Encephalographic (MEG) Corresponding to Subjective Preference

In order to attain further knowledge of brain activities, responses on EEG corresponding to subjective preference have been conducted including in change of reverberation time (T_{sub}) with continuous test signals that could not be performed by the evoked potentials applying short signals less than 0.9 s. First of all, to find a distinctive feature through EEG changing the delay time of a single reflection (Δt_1) to reconfirm the SVR results. In order to obtain individual differences clearly, further investigation was conducted in response to MEG following changes to Δt_1. Then, responses of EEG in changes to T_{sub} are discussed. Effects of the typical spatial factor (IACC) are investigated by recording EEG.

4.3.1 EEG in Response to Change of Δt_1

In this experiment, music motif B (Arnold's Sinfonietta, Opus 48, a 5 s segment of the 3rd movement) was selected as the sound source (Burd 1969; Ando 1985). The delay time of a single reflection Δt_1 was alternatively adjusted to 35 ms (a preferred condition) and 245 ms (a condition of echo disturbance). The EEG of ten pairs from T_3 and T_4 was recorded for about 140 s, and experiments were repeated over a total of 3 days. Eleven, 22–26-year-old subjects participated in the experiment. The subject was asked to close his eyes while listening to the music during the recording of the EEG. Two loudspeakers were arranged in front of the subject. Thus, the IACC was kept at a constant value of near unity. The sound pressure level was fixed at 70 dBA peak, in which the amplitude of the single reflection was the same as that of the direct sound, $A_0 = A_1 = 1$. The leading edge of each sound signal was recorded at the same time for analysis of the EEG. The EEG recorded was sampled at greater than 100 Hz after passing through a filter width of 5–40 Hz with a slope of 140 dB/octave.

In order to find the brain activity corresponding to subjective preference, an attempt was made to analyze the effective duration of the ACF, τ_e in the α-wave range (8–13 Hz) of the EEG. First considering that the subjective preference judgment needs at least 2 s to develop a psychological present, the running integration interval ($2T$) was examined for periods between 1.0 and 4.0 s.

① A satisfactory duration $2T = 2$–3 s in the ACF analysis was found only from the left hemisphere, but not from the right (Ando and Chen 1996).
Table 4.2 indicates the results of the analysis of the variance for values of τ_e of the α-wave obtained at $2T = 2.5$ s. Though the individual difference is significant ($p < 0.01$), the factor Δt_1 (LR: $p < 0.025$) is significant. However, it is significant for an interference effect between factors Δt_1 and LR ($p < 0.01$).

Table 4.2 Results of the analysis of variance for τ_e of ACF of α-wave, with changes in Δt_1

Factor	F	Significance level
Subject	93.1	<0.01
Hemisphere, LR	1.0	
Delay time, Δt_1	5.8	<0.025
Subject and LR	8.9	<0.01
Subject and Δt_1	0.4	
LR and Δt_1	9.6	<0.01
Subject, LR, and Δt_1	0.4	

Therefore, in order to analyze the data in detail for each category Δt_1 and LR, we show the averaged value of τ_e in the α-wave with 11 subjects in Fig. 4.6. A clear tendency is apparent.

② Values of τ_e at Δt_1 = 35 ms are significantly longer than those at Δt_1 = 245 ms ($p < 0.01$) only on the left hemisphere, not on the right. Ratios of τ_e values in the α-wave range at Δt_1 = 35 and 245 ms, for each subject, are shown in Fig. 4.7. Remarkably, all individual data indicate that the ratios in the left hemisphere at the preferred condition of 35 ms are longer than those in the right hemisphere.

③ Thus, the results reconfirm that, when the Δt_1 is changed, the left hemisphere is highly activated, and the value of τ_e for the α-wave of this hemisphere corresponds well with the subjective preference. The α-wave has the longest period in the EEG in the awaking stage and may indicate feelings of "pleasantness" and "comfort", a preferred condition, which is widely accepted. Thus, a long value of τ_e in the α-wave may relate to the long N_2 latency of SVR at the preferred condition as shown in Figs. 4.4 and 4.5.

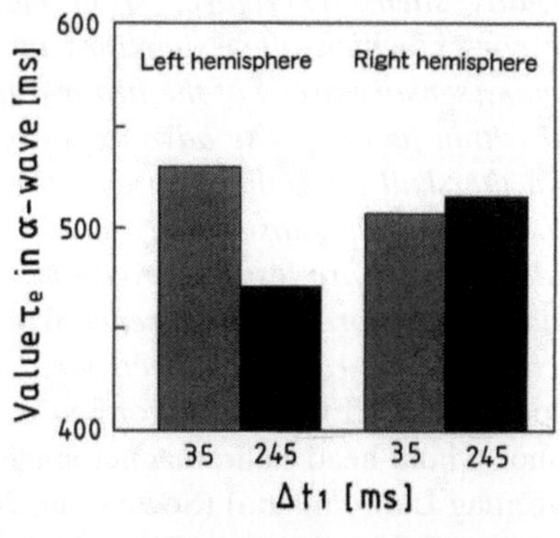

Fig. 4.6 Averaged ACF τ_e values of the EEG-alpha wave in change of Δt_1: 35 and 245 ms (11 subjects). *Left* Left hemisphere. *Right* Right hemisphere. Significant difference in ACF τ_e values may be found on the left hemisphere, but it does not on the right

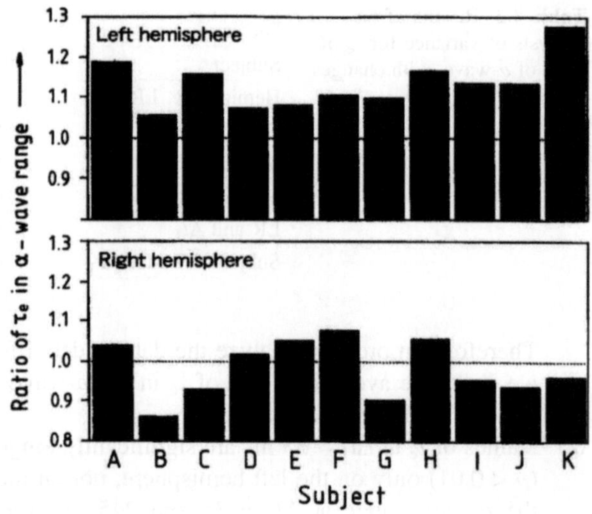

Fig. 4.7 Ratios of ACF τ_e values of the EEG-alpha wave of 35 and 245 ms in change of Δt_1 for each individual subject, *A–K* [τ_e value at 35 ms]/[τ_e value at 245 ms]. *Above* Left hemisphere. *Below* Right hemisphere. Ratios of ACF τ_e values of the EEG-alpha wave are always greater on the left hemisphere than on the right

4.3.2 MEG in Response to Change of Δt₁

In MEG studies, the weak magnetic fields produced by electric currents flowing in neurons are measured with multiple channel SQUID (superconducting quantum interference device) gradiometers; which enable the study of many interesting properties of the working human brain. MEG accurately detects superficial tangential currents, whereas EEG is sensitive to both radial and tangential current sources and also reflects activity in the deepest parts of the brain. Only currents that have a component tangential to the surface of a spherically symmetric conductor produce a sufficiently strong magnetic field outside of the brain; radial sources are thus externally silent. Therefore, MEG measures mainly, activity from the fissures of the cortex, which often simplifies interpretation of the data. Fortunately, all the primary sensory areas of the brain—auditory, somatosensory, and visual—are located within fissures. The advantages of MEG over EEG result mainly from the fact that the skull and other extracerebral tissues are practically transparent to magnetic fields, but substantially alter the current flow. Thus, magnetic patterns outside the head are less distorted than the electrical potentials on the scalp. Further, magnetic recording is reference-free, whereas electric brain maps depend on the location of the reference electrode.

Measurements of responses on MEG were performed in a magnetically shielded room using a 122-channel whole-head neuromagnetometer as shown in Fig. 4.8 (Neuromag-122^{TM}, Neuromag Ltd., Finland) (Soeta et al. 2002). The source signal was the word, "piano," with a 0.35 s duration. The minimum value of the moving τ_e, i.e., $(\tau_e)_{min}$, was about 20 ms. It is worth noticing that this value is close to the most preferred delay time of the first reflection of sound fields with continuous speech (Ando and Kageyama 1977). In the present experiment, the delay time of

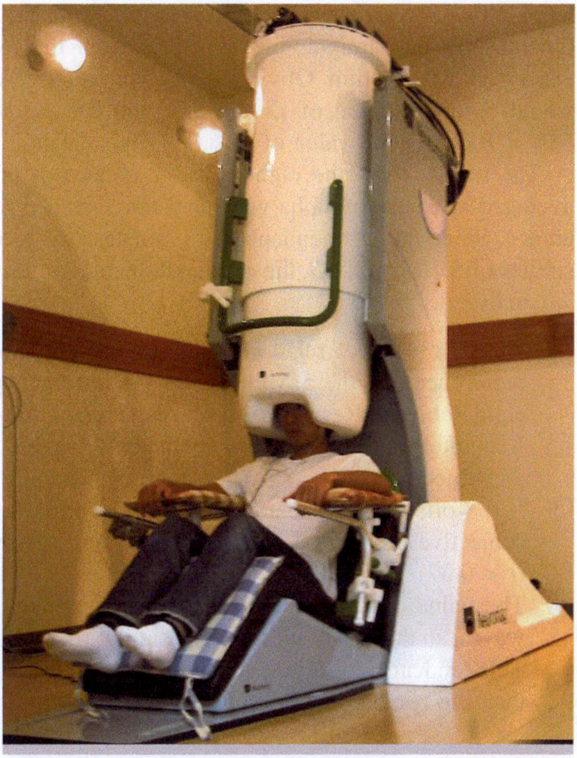

Fig. 4.8 Magnetometer used in recording the magnetoencephalography (MEG)

the single reflection (Δt_1) was set at five levels (0, 5, 20, 60, and 100 ms). The direct sound and a single reflection were mixed and the amplitude of the reflection was the same as that of the direct sound ($A_0 = A_1 = 1$). The auditory stimuli were binaurally delivered through plastic tubes and earpieces into the ear canals without any metric material. The sound pressure level, which was measured at the end of the tubes, was fixed at 70 dBA.

Seven, 23–25-year-old subjects participated in the experiment. All had normal hearing. In accordance with the PCT, each subject compared ten possible pairs per session, and a total of ten sessions were conducted for each subject. Measurements of magnetic responses were performed in a magnetically shielded room. Similar to the above EEG measurements, the paired-auditory stimuli were presented in the same way as in the subjective preference test. During measurements, the subjects sat in a chair with their eyes closed. To compare the results of the MEG measurements with the scale values of the subjective preference, combinations of a reference stimulus (Δt_1 = 0 ms) and test stimuli (Δt_1 = 0, 5, 20, 60, and 100 ms) were presented alternately 50 times, and the MEGs were analyzed. The magnetic data was recorded continuously with a filter of 0.1–30.0 Hz and digitized with a sampling rate of 100 Hz. Eight channels that had larger amplitude of N1m response in each hemisphere were selected for the ACF analysis. We analyzed the

MEG-alpha wave for each of the paired stimuli, for each subject. The value of τ_e obtained by the straight line for 5 dB from the top of the normalized ACF is expressed in logarithm. Obviously, for the preferred condition at $\Delta t_1 = 5$ ms of the sound field the value of $\tau_e \approx 0.5$ s, and for the condition of echo disturbance ($\Delta t_1 = 100$ ms), the value of $\tau_e \approx 0.3$ s is smaller.

The results from the eight subjects confirm a linear relationship between the averaged τ_e values of alpha wave and the averaged scale values of subjective preference. Since the left hemisphere dominates Δt_1, reconfirming the aforementioned studies of SVR and EEG, the results of the individual level from the left hemisphere are analyzed.

① An almost direct relationship between individual scale values of subjective preference and the τ_e values over the left hemisphere as found in each of the eight subjects. Results for each of the eight subjects are shown in Fig. 4.9.
② Remarkably, the correlation coefficient, r, was achieved by more than 0.94 for all subjects.
③ It is worth noticing that there is little relationship between the scale values of subjective preference and the amplitude of α wave, $\Phi(0)$, in both hemispheres ($r < 0.37$).
④ The value of τ_e is the degree of similar repetitive features included in alpha waves, so that the brain repeats a similar rhythm under the preferred conditions. This tendency for a larger τ_e under the preferred condition is more significant than the results on EEG alpha waves as mentioned above.

4.3.3 EEG in Response to Change of T_{sub}

Now let us examine the values of τ_e in the α-wave with changes to the subsequent reverberation time (T_{sub}) relative to the scale values of subjective preference. Ten student subjects participated in the experiment (Chen and Ando 1996). The sound source used was music motif B, as described with $(\tau_e)_{min} \sim 40$ ms, so that the most preferred reverberation time calculated is $(T_{sub})_p \sim 23\ (\tau_e)_{min} = 0.92$ s (Sect. 6.3). Ten, 25–33-year-old subjects participated in the experiment. The EEG from the left and right hemispheres was recorded. Values of τ_e of the α-wave, for the duration, $2T = 2.5$ s, were also analyzed here.

First consider the averaged values of τ_e of the α wave, shown in Fig. 4.10.

① Clearly, the values of τ_e are longer at close to the preferred condition 0.92 s, $T_{sub} = 1.2$ s than those at $T_{sub} = 0.2$ s in the left hemisphere, while the values of τ_e are longer at $T_{sub} = 1.2$ s than those at $T_{sub} = 6.4$ s.
② It is significant in the left hemisphere. However, this is not true for the right hemisphere.

The results of analysis of the variance are that although there is a large individual difference, a significant difference is achieved for T_{sub} in the pair of 0.2 and 1.2 s

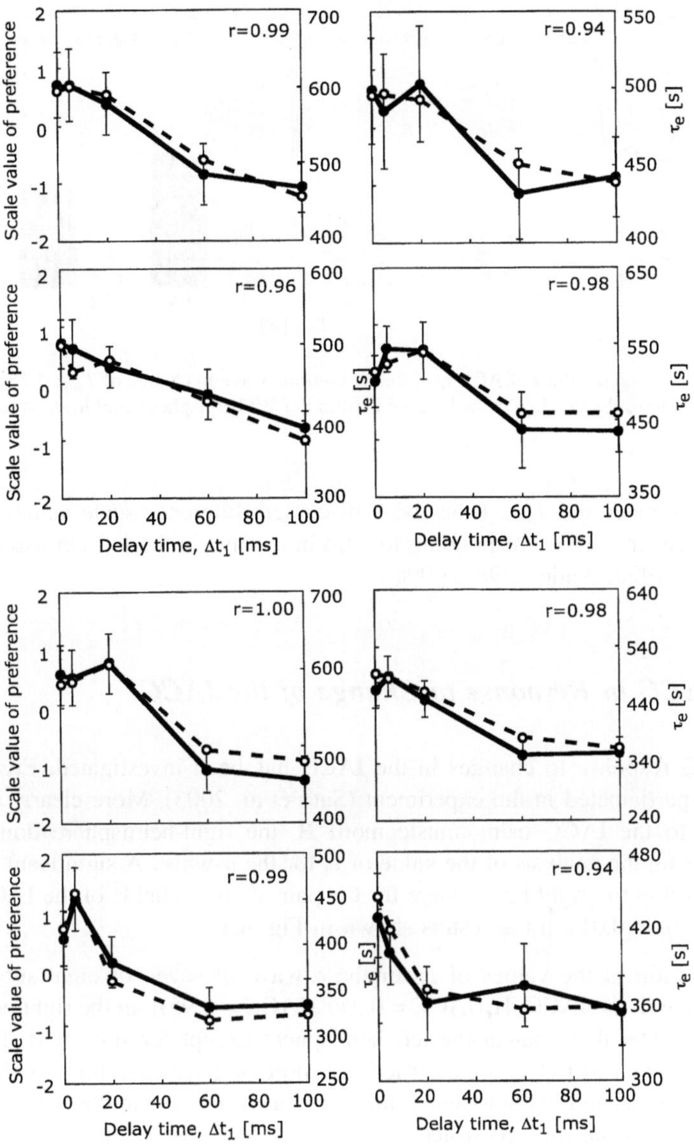

Fig. 4.9 Good correspondence between the scale value of subjective preference and the averaged ACF τ_e value of the MEG-alpha wave over the left hemisphere. (eight subjects). The averaged τ_e value and the scale value are the highest correlations over the eight channels. *Open circles* Scale values of subjective preference; *filled circles* averaged τ_e values of MEG alpha wave, error bars being standard errors

($p < 0.05$), and interference effects are observed for factors Subject and LR ($p < 0.01$), and LR and T_{sub} ($p < 0.01$). No such significant differences are achieved for the pair at 1.2 and 6.4 s, but there are interference effects between Subject and

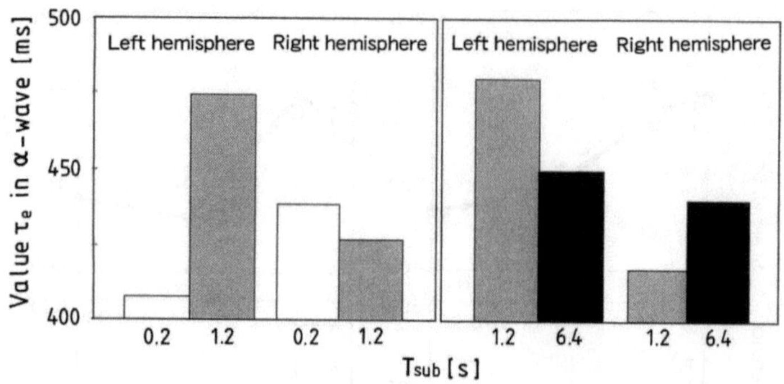

Fig. 4.10 Averaged values of ACF τ_e of the EEG-alpha wave in change of T_{sub}: 0.2 and 1.2 s; 1.2 and 6.4 s for ten subjects. *Left* in each figure is from the left hemisphere, and *Right* is from the right hemisphere

LR, and Subject and T_{sub}. It has been discussed difference scale values of preference, however, well corresponding to ratio in α-wave in the left hemisphere (Chen and Ando 1996; Ando 1998, 2009a).

4.3.4 EEG in Response to Change of the IACC

The EEG response to changes in the IACC has been investigated. Eight student subjects participated in the experiment (Sato et al. 2003). More clearly here, with changes to the IACC using music motif B, the right-hemisphere dominance is shown using the analysis of the value of τ_e for the α-wave. A significant difference is achieved in the right hemisphere for the pair of sound fields of the IACC = 0.95 and 0.30 ($p < 0.01$) in the results shown in Fig. 4.11.

① The ratio of the values of τ_e for the α-wave of seven of eight subjects with change in the IACC, [τ_e(IACC = 0.3)/τ_e(IACC = 0.95)], in the right hemisphere are greater than that in the left hemisphere except for subject B (Fig. 4.12). Thus, as far as IACC is concerned, the more preferred condition with a smaller IACC is related to the longer value of τ_e for the α-wave in the right hemisphere in most of the subjects tested.

② Also, it has been reconfirmed by experiments on MEG with the same speech signal in change of the IACC (0.27, 0.61 and 0.90) that the values of τ_e and the maximum amplitude of the CCF were increased when decreasing the IACC (Soeta et al. 2005).

Table 4.1 summarizes hemisphere dominance obtained by analysis of τ_e in α-waves, with changes in LL, Δt_1, T_{sub}, and the IACC. This finding may suggest that the value of τ_e of the α-waves is an objective index for designing preferred conditions of the human environment.

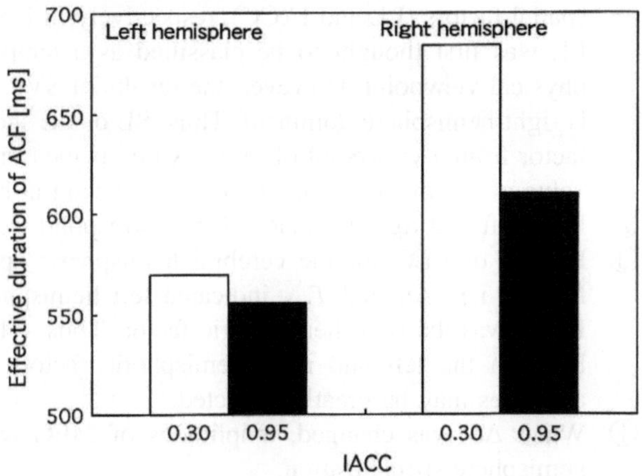

Fig. 4.11 Averaged values of ACF τ_e of the EEG-alpha wave in change of the IACC for the pair of IACC = 0.30 and 0.95. *Left* in the figure is from the left hemisphere, and *Right* is from the right hemisphere

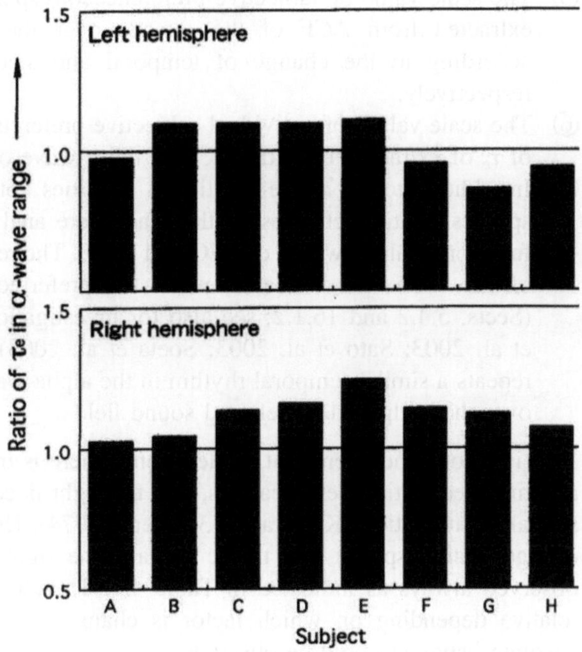

Fig. 4.12 Ratio of ACF τ_e values of the EEG-alpha wave from the left hemisphere (T_3) and the right hemisphere (T_4) for each of eight subjects, *A–H* [τ_e value at IACC = 0.30]/[τ_e value at IACC = 0.95]. Ratio of ACF τ_e values are greater on the right hemisphere than those of the left except for subject B

4.4 Specialization of Cerebral Hemispheres for Temporal and Spatial Factors of the Sound Field

Recordings over the left and right hemispheres of ABR, SVR, EEG, and MEG revealed the following evidence as summarized in Table 4.1 *(Ando 2011).*

① The left and right amplitudes of the early SVR, A(P_1–N_1) indicate that the left and right hemispheric dominance are due to the temporal factor (Δt_1) and spatial factors (LL and IACC), respectively. It is worth noticing that the SL or LL was first thought to be classified as a temporal-monaural factor from a physical viewpoint. However, the results of SVR indicate that the sound level is right hemisphere dominant. Thus, SL or LL should be classified as a spatial factor from a viewpoint of brain, which is measured by the geometric average value of the binaural sound energies arriving at both ears.

② Both left and right latencies of N_2 correspond well to the values of IACC.

③ Results of EEG for the cerebral hemispheric specialization of the temporal factors, i.e., Δt_1 and T_{sub} indicated left hemisphere dominance, whereas the IACC was the right hemispheric factor. Thus, a high degree of independence between the left and right hemispheric factors in judgments of subjective attributes may be greatly expected.

④ When Δt_1 was changed, amplitudes of MEG recorded reconfirmed the left hemisphere specialization.

⑤ The scale value of subjective preferencecorresponds well to the value of τ_e of extracted from ACF of the α-wave over the left and right hemispheres according to the change of temporal and spatial factors of sound fields, respectively.

⑥ The scale values of individual subjective preference relate directly to the value of τ_e of extracted from the ACF of the α-wave of MEG.

⑦ In addition to the above-mentioned activities both on the left and right hemispheres, spatial activities on the brain were analyzed by the cross-correlation function of alpha waves of EEG and MEG. The results showed that a large area of the brain was activated, when the preferred sound fields are presented (Sects. 5.4.2 and 16.1.2; see also for investigations in visual fields: Okamoto et al. 2003; Sato et al. 2003; Soeta et al. 2003). These imply that the brain repeats a similar temporal rhythm in the alpha-wave range throughout the area over the scalp under preferred sound fields.

It is reconfirmed here that the left hemisphere is mainly associated with speech and time-sequential identifications, and the right is concerned with nonverbal and spatial identification (Kimura 1973; Sperry 1974). However, when the IACC was changed using speech and music signals, the right hemisphere dominance was observed always as indicated in Table 4.1. Therefore, hemispheric dominance is relative depending on which factor is changed in the comparison pair and no absolute behavior could be observed.

It has been discovered that the listening level (LL) and the IACC are dominantly associated with the right cerebral hemisphere, and the temporal factors, Δt_1 and T_{sub}, the sound field in a room are associated with the left.

It is remarkable that, for example, "cocktail party effects" could well be explained in terms of specialization of the human brain, because speech signal is processed in the left hemisphere, and independently the directional information of a target speaker is mainly processed in the right hemisphere.

4.5 Model of Auditory-Brain System

Based on the above-mentioned physical system and physiological responses, a model of the central auditory signal processing model (Ando 1985) has been reconfirmed as shown in Fig. 4.13. *Applying this model, a wide range of research works may be available. For example, automatic speech recognition in telephone and in rooms of more than 7,000 existing languages in the world are being initiated (Ando 2015). Classification of music by values of effective duration of ACF* $(\tau_e)_{min}$ *by use of musical notes for selecting concert programs blending temporal factors of the sound field in an enclosure to be performed (Ando 2009), and automatic noise measurement with classifying sources, for example, aircraft noise or traffic noise, after identification of noise source by use of ACF factors (Soeta and Ando 2015), and effects of environmental noise on developing brain-specialization and accumulated effects of noise on mind (Ando 2011).*

The model consists of the autocorrelation mechanisms, the interaural cross-correlation mechanism between the two auditory pathways, and the specialization of human cerebral hemispheres for temporal and spatial factors of the sound field. In addition, according to the relationship between temporal and spatial percepts of sound as well as subjective preference of the sound field and physiological phenomena in changes with variation to the acoustic factors, a model is reconfirmed and formed as shown in Fig. 4.13 (Ando 1985). In this figure, a sound source $p(t)$ is located at r_0 in a three-dimensional space and a listener is sitting at r which is defined by the location of the center of the head, $h_{l,r}(r|r_0, t)$ being the impulse responses between r_0 and the left and right ear canal entrances. The impulse responses of the external ear canal and the bone chain are $e_{l,r}(t)$ and $c_{l,r}(t)$, respectively. The velocity of the basilar membrane is expressed by $V_{l,r}(x, \omega)$, x being the position along the membrane.

Amplitudes of waves (I–IV) of the ABR reflect to the sound pressure levels as a function of the horizontal angle of incidence to a listener (Sect. 4.1.1). Such neural activities, in turn, include sufficient information to attain the ACF, probably at the lateral lemniscus as indicated by $\Phi_{ll}(\sigma)$ and $\Phi_{rr}(\sigma)$. In fact, the time domain analysis of firing rate from auditory nerve of cat reveals a pattern of ACF rather than the frequency domain analysis (Secker-Walker and Searle 1990). Pooled inter-spike interval distributions resemble the short time or the running ACF for the low-frequency component. And, the pooled interval distributions for sound stimuli consisting of the high-frequency component resemble the envelope to running ACF as shown in Fig. 4.1 (Cariani and Delgutte 1996a). From a viewpoint of the missing fundamental or pitch of complex components judged by humans, the running ACF must be processed in frequency components below about 5 kHz and the fundamental frequency below 1,200 Hz (Inoue et al. 2001).

As is also discussed, the neural activity (wave V together with waves IV_l and IV_r) may correspond to the IACC as shown in Figs. 4.2 and 4.3. Thus, the interaural cross-correlation mechanism exists at and around the inferior colliculus. It is concluded that the output signal of the interaural cross-correlation mechanism

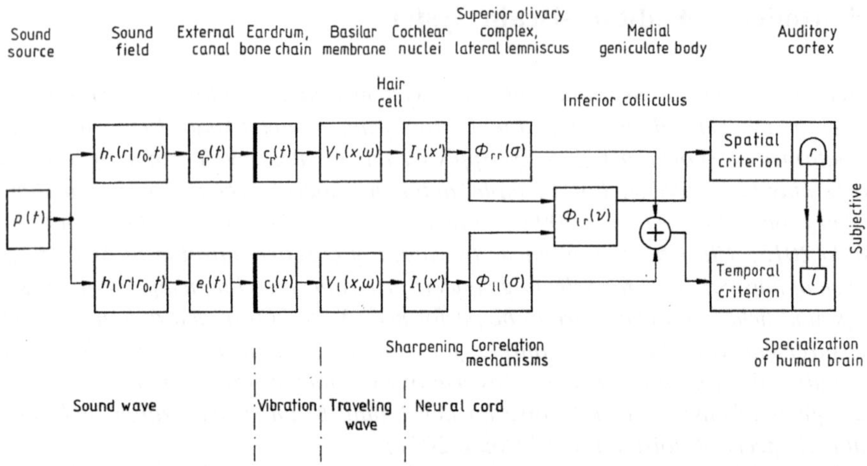

Fig. 4.13 Central auditory signal processing model for subjective responses. $p(t)$ Source sound signal in the time domain; $h_{l,r}(r/r_0, t)$ head-related impulse responses from a source position of r_0 to the left and right ear entrances of a listener at r; $e_{l,r}(t)$ impulse responses of left and right external canals, from the left and right ear entrances to the left and right eardrums; $c_{l,r}(t)$ impulse responses for vibration of left and right bone chains, from the eardrums to oval windows, including transformation factors into vibration motion at the eardrums: $V_{l,r}(x, \omega)$ travelling wave forms on the basilar membranes, where x the position along the left and right basilar membrane measured from the oval window; $I_{l,r}(x')$ sharpening in the cochlear nuclei corresponding to roughly the power spectra of input sound, i.e., responses of a single pure tone ω tend to a limited region of nuclei. These neural activities may be enough to convert into activities similar to the ACF; $\Phi_{ll}(\sigma)$ and $\Phi_{rr}(\sigma)$ ACF mechanisms in the left and right auditory pathways, respectively. Symbol $\oplus$ signifies that signals are combined; $\Phi_{lr}(\nu)$ IACF mechanism (Ando 1985); r and l: specialization for temporal and spatial factors of the left and right human cerebral hemispheres, respectively. Temporal sensations (Sect. 5.1) and spatial sensations (Sect. 5.2) may be processed in the left and right hemispheres according to the temporal factors extracted from the ACF and the spatial factors extracted from the IACF, respectively. The overall subjective preference and annoyance may process in both hemispheres in relation to the temporal and spatial factors (Ando 2002, 2009a)

including the IACC may be dominantly connected to the right hemisphere. Also, the sound pressure level may be expressed by a geometrical average of ACFs for the two ears at the origin of time ($\sigma = 0$) and in fact appears in the latency at the inferior colliculus, may be processed in the right hemisphere.

Based on the model, we well describe temporal and spatial percepts (Ando et al. 1999; Ando 2002), in turn, any subjective attributes of sound fields in terms of processes in the auditory pathways and the specialization of two cerebral hemispheres.

Chapter 5
Temporal and Spatial Primary Percepts of the Sound and the Sound Field

We shall discuss temporal percepts consisting of four such that pitch, loudness, duration, and timbre in relation to the temporal factors of the sound. Three spatial percepts, i.e., localization including movement of sound source on the stage, apparent source width (ASW), and subjective diffuseness are well described in relation to the spatial factors of the sound field.

5.1 Temporal Percepts in Relation to the Temporal Factors of the Sound

5.1.1 Pitches of Complex Tones

We shall show that a factor extracted from the autocorrelation function (ACF) of sound signal directly describes the pitch perception.

What is interesting is that harmonic complexes having no energy at the fundamental frequency in their power spectra still can produce strong "low" pitches at the fundamental itself. It is thus the cases for complex tones with a "missing fundamental" that strong pitches are heard that correspond to no individual frequency component, and this raises deep questions about whether patterns of pitch perception are consistent with frequency-domain representations.

A pitch-matching test comparing pitches of pure and complex tones was performed to reconfirm previous results (Sumioka and Ando 1996). The test signals were all complex tones consisting of harmonics 3–7 of a 200 Hz fundamental. All tone components had the same amplitudes, as shown in Fig. 5.1. As test signals, the two waveforms of complex tones, (a) in-phases and (b) random phases, were applied as shown in Fig. 5.2. Starting phases of all components of the in-phase stimuli were set at zero. The phases of the components of random-phase stimuli were randomly set to avoid any periodic peaks in the real waveforms. As shown in Fig. 5.3, the normalized ACF of these stimuli were calculated at the integrated interval $2T = 0.8$ s. Though the waveforms differ greatly from each other, as shown in Fig. 5.2, thus their ACF are experimentally and theoretically identical. The time

Y. Ando, *Opera House Acoustics Based on Subjective Preference Theory*,
Mathematics for Industry 12, DOI 10.1007/978-4-431-55423-3_5

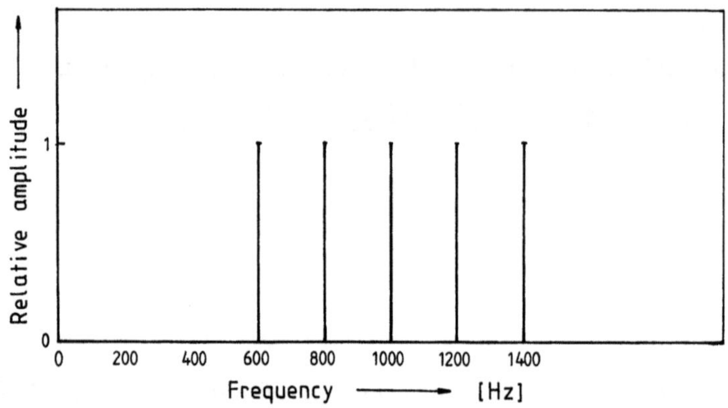

Fig. 5.1 Complex tones tested with five pure-tone components of identical amplitude of 600, 800, 1000, 1200, and 1400 Hz

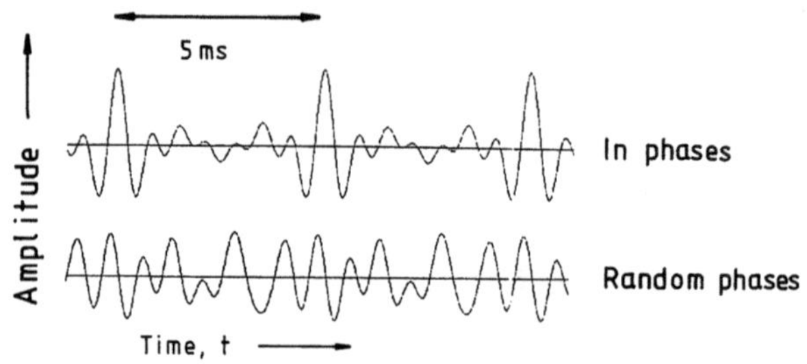

Fig. 5.2 Waveforms of the complex tone with the five pure-tone components in-phases and random phases

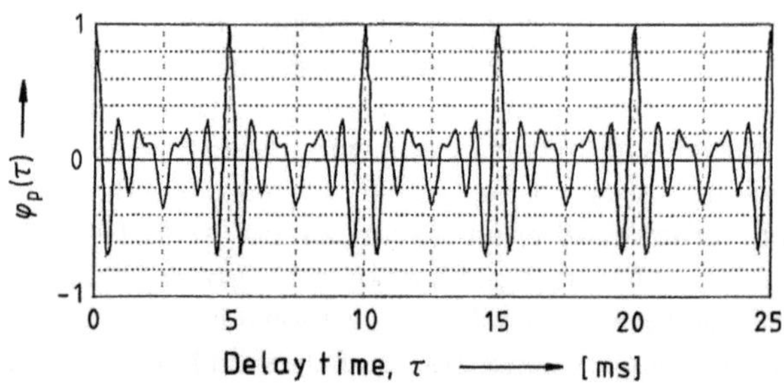

Fig. 5.3 Normalized ACF analyzed for the complex tone with the five pure-tone components both in-phases and random phases. It has been known mathematically that it is identical with any phase condition

delay at the first maximum peak of the ACF, τ_1 equals to 5 ms (200 Hz) corresponding well to the fundamental frequency. The subjects were five musicians (two male and three female, 20–26 years of age). Test signals were produced from the loudspeaker in front of each subject in a semi-anechoic chamber. The sound pressure level (SPL) of each complex tone at the center position of the listener's head was fixed at 74 dB by analysis of the ACF $\Phi(0)$. The distance between a subject and the loudspeaker was 0.8 m ± 1 cm.

Probability of matching frequencies counted for each 1/12-octave band (chromatic scale) of the in-phase stimuli and random-phase stimuli is shown in Fig. 5.4. The dominant pitch of 200 Hz is included neither in the spectrum nor in the real waveform of random phases. But it is obviously included in the period in the ACF. For both in-phase and random-phase conditions, about 60 % of the responses clustered within a semitone of the fundamental. Results obtained for pitch under the

Fig. 5.4 Probability obtained by the pitch-matching tests for five subjects adjusted

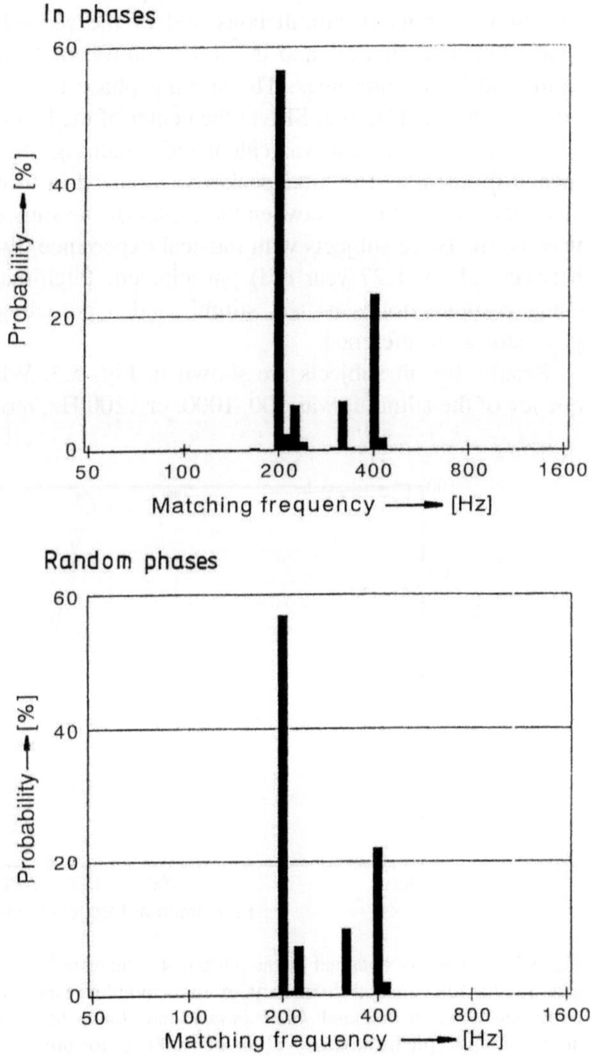

two conditions are definitely similar. In fact, the pitch strength remains the same under both conditions. Thus, pitch of complex tones can be predicted from the time delay at the first maximum peak of the ACF, τ_1. This result confirmed the result obtained by Yost (1996) who demonstrated that pitch perception of iterated rippled noise is greatly affected by the first major ACF peak of the stimulus signal.

5.1.2 Frequency Limits of the ACF Model

For fundamental frequencies of 500, 1000, 1200, 1600, 2000, and 3000 Hz, stimuli consisting of two or three pure-tone components were produced (Inoue et al. 2001). The two-component stimuli consisted of the second and third harmonics of the fundamental frequency, and the three-component stimuli consisted of the second, third, and forth harmonics. The starting phase of all components was adjusted to zero (in-phase). The total SPL at the center of the listener's head was fixed at 74 dB. The ACF of all stimuli was calculated obtaining the peak τ_1 related to the fundamental frequency. The loudspeaker was placed in front of a subject in an anechoic chamber. The distance between the center of the subject's head and the loudspeaker was 0.8 m. Three subjects with musical experience (two male and one female, aged between 21- and 27-year old) participated. Pitch-matching tests were conducted using complex tones as test stimuli and a pure tone generated by a sinusoidal generator as a reference.

Results for all subjects are shown in Fig. 5.5. Whenever the fundamental frequency of the stimulus was 500, 1000, or 1200 Hz, more than 90 % of the responses

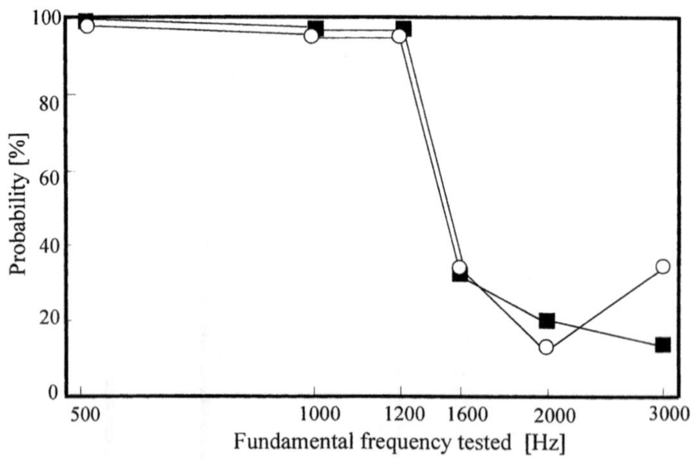

Fig. 5.5 Probability obtained by the pitch-matching tests for three subjects adjusted to a pure tone near to the fundamental frequency of two complex tones. *Open circles* are results for two frequency components, and *filled squares* are those for three components. This shows the limitation up to pitch frequency of about 1,200 Hz for the ACF model

were obtained from all subjects under both conditions clustered around the fundamental frequency. When the fundamental frequencies of the stimuli were 1600, 2000, or 3000 Hz, however, the probability, that the subjects adjusted the frequency of the pure tone to the calculated fundamental frequency, is extremely decreased. These results imply that the ACF model is applicable when the fundamental frequency of stimuli is below 1,200 Hz.

5.1.3 Loudness of Sharply Filtered Noise

In this section, we shall show that loudness of sharply filtered noise with identical SPL is not constant even within the critical band. And, loudness of the pure tone was significantly larger than that of sharply filtered noises, and loudness increased with the increasing τ_e extracted from ACF within the critical band.

The bandwidth (Δf) of a sharp filter was changed by using the cutoff slope of 2,068 dB/octave, which was realized by a combination of two filters (Sato et al. 2002). Factors of τ_1, τ_e, and ϕ_1 are extracted from the ACF. In fact, the filter bandwidth of 0 Hz included only its slope component. All source signals were the same SPL at 74 dBA, which was accurately adjusted by measurement of the ACF at the origin of the delay time, $\Phi(0)$.

The loudness judgment was performed by the paired-comparison test (PCT) for which the ACF of the band-pass noise was changed. A headphone delivered the same sound signal to the two ears. Thus, the IACC was kept constant at nearly unity. Sound signals were digitized at a sampling frequency of 48 kHz. Five subjects with normal hearing were seated in an anechoic chamber and asked to judge which of the two paired sound signals were perceived to be louder. Stimulus durations were 1.0 s, rise and fall times were 50 ms, and silent intervals between the stimuli were 0.5 s. A silent interval of 3.0 s separated each pair of stimuli, and the pairs were presented in random order.

Fifty responses (5 subjects × 10 sessions) to each stimulus were obtained. Consistency tests indicated that all subjects had a significant ($p < 0.05$) ability to discriminate loudness. The test of agreement also indicated that there was significant ($p < 0.05$) agreement among all subjects. A scale value (SV) of loudness was obtained by applying the law of comparative judgment (Thurstone's case V) and was confirmed by goodness of fit.

The relationship between the SV of loudness and the filter bandwidth is shown in Fig. 5.6a–c. The SV difference of 1.0 corresponds about 1 dB due to the preliminary experiment. For all three-center frequencies (250, 500, 1000 Hz), the SV of loudness is maximal for the pure tone with the infinite value of τ_e and large bandwidths with minima at smaller bandwidths (40, 80, 160 Hz, respectively). From the dependence of τ_e on filter bandwidth, we found that loudness increases with increasing τ_e within the "critical bandwidth". Results of analysis of variance (ANOVA) for the SV of loudness are indicated that for all center frequencies tested, the SV of loudness of pure tone was significantly larger than that of other band-pass

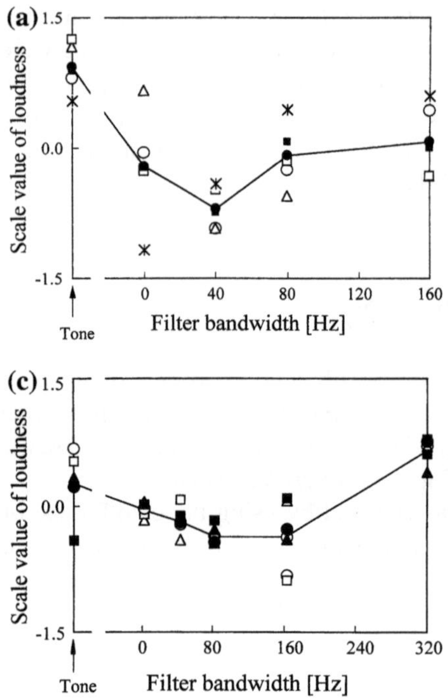

Fig. 5.6 Scale values (SV) of loudness obtained by the PCT as a function of the bandwidth of noise by applying sharply filters with the cutoff slope of 2,068 dB/octave. *Different symbols* indicate the SV obtained with different subjects. **a** f_c = 250 Hz. **b** f_c = 500 Hz. **c** f_c = 1,000 Hz

noises within the critical band ($p < 0.01$). Consequently, loudness of the band-pass noise with identical SPL was not constant within the critical band. Also, loudness of the pure tone was significantly larger than that of sharply filtered noises, and loudness increased with increasing τ_e within the critical band.

5.1.4 Duration Sensation

As a fourth of the temporal and primary percept, we investigated the duration sensation by the PCT. As is known that duration is indicated in a musical note, results show that relatively shorter duration sensation is observed for higher pure tone (3,000 Hz) than that of 500 Hz. And, duration sensation of 500 Hz pure tone is similar to that of complex tone with 500 Hz pitch.

Experiments for pure and complex tones were performed by the PCT (Saifuddin et al. 2002). Throughout this investigation, the SPL was fixed at 80 dBA. Waveform amplitudes during stimulus onsets and offsets were ramped with rise/fall times of 1 ms for all stimuli tested, the time required to reach a threshold 3 dB below the steady level. Perceived durations of two-component complex tones (3,000 and

3,500 Hz) having a fundamental at 500 Hz were compared with those evoked by pure-tone stimuli at 500 and 3,000 Hz. Pairs consisting of two stimuli were presented randomly to obtain SV for duration sensation (DS). Three signal durations, including rise/fall segments, were used for each of the stimuli: D = 140, 150, and 160 ms. There were thus 9 stimulus conditions, and 36 pair-wise stimulus combinations. The source stimuli were presented in a darkened soundproof chamber from a single loudspeaker at the horizontal distance of 74 (±1) cm from the center of the seated listener's head. Ten students participated in both experiments as subjects of normal hearing levels (22 and 36 years old). Each pair of stimuli was presented five times randomly within every session for each subject.

Observed SV for the perceived durations of the 9 stimuli are shown in Fig. 5.7. While signal duration and stimulus periodicity had major effects on perceived duration, the number of frequency components (1 vs. 2) did not. Perceived durations of tones with the same periodicity (f = 500 Hz and F_0 = 500 Hz) were almost identical, while durations for pure tones of different frequencies (f = 500 Hz vs. f = 3,000 Hz) differed significantly, by approximately 10 ms (judging from equivalent SV, the 500 Hz pure tone appeared ~ 10 ms longer than the 3,000 Hz tone). Thus, the duration (DS) of the higher frequency pure tone (3,000 Hz; τ_1 = 0.33 ms) was found to be significantly shorter ($p < 0.01$) than that of either the pure tone (frequency: 500 Hz; τ_1 = 2 ms) or the complex tone (fundamental frequency: F_0 = 500 Hz; τ_1 = 2 ms). Also, the SV of DS between the two pure tones: τ_1 = 2 (500 Hz) and 0.33 ms (3,000 Hz) are almost parallel, so that the effects of periodicity (τ_1) and signal duration (D) on the apparent duration (DS) are independent and additive. Therefore, for these experimental conditions, we may express tentatively

$$S_L = f(\tau_1, D) = f(\tau_1) + f(D) \tag{5.1}$$

where τ_1 is extracted from the stimulus ACF. But, it is recommended further investigations on duration for wider range of pitches of complex and pure tones.

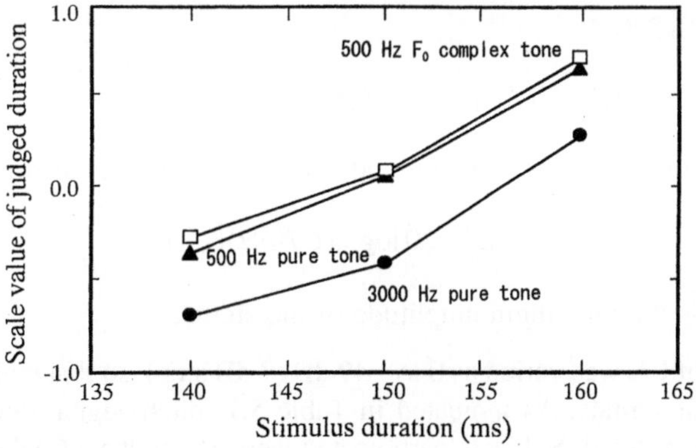

Fig. 5.7 SV of duration sensation (DS) obtained by the PCT. *Open squares* Complex tone (F_0 = 500 Hz) with pure-tone components of 3,000 and 3,500 Hz; *filled triangles* 500-Hz pure tone; *filled circles* 3,000-Hz pure tone

5.1.5 Timbre of an Electric Guitar Sound with Distortion

Timbre is defined as an aspect of sound quality independent of loudness, pitch, and duration. The quality of sound texture that distinguishes two notes of equal pitch, loudness, and duration that are played by different musical instruments. An attempt is made here to investigate the relationship between the temporal factors extracted from the ACF of an electric guitar sound and dissimilarity representing the difference of timbre with a difference of distortion.

As shown in Fig. 2.1, a factor $W_{\phi(0)}$ is defined by the delay time at the first 0.5 crossing of the normalized ACF, $\phi(\tau)$ that deeply related to a global frequency component of the source signal. This value is equivalent to the factor W_{IACC} extracted from the IACF.

An electric guitar with the "distortion" is a main instrument in pops and rock music. Previously, Marui and Martens (2005) investigated timbre variations by the use of three types of nonlinear distortion processors with differing level of Zwicker Sharpness (Zwicker and Fastl 1999). Resulting dissimilarity data revealed three dimensions, one of which distinguished between effect types, while other two of which were highly correlated with ratings on adjective scales anchored by the pairs dark–bright and sharp–dull. But no attempt has been made in relation to the objective parameters for describing dissimilarity data.

In this study, we examine whether or not timbre is described by the temporal factors extracted from the running ACF of the source signal.

Experiment 1

The purpose of this experiment is to find the factor extracted from the running ACF contributing to the dissimilarity of sounds changing the strength of distortion by the use of computer. The distortion of music signal $p(t)$ was processed by a computer program, such that: when $|p(t)| \leq C$

$$p(t) = p(t), \tag{5.2}$$

and when $|p(t)| \geq C$

$$p(t) = +C,\ p(t) \geq C;\ p(t) = -C,\ p(t) \leq -C, \tag{5.3}$$

where C is the cut-off pressure amplitude, and its level is defined by

$$\mathrm{CL} = 20 \log_{10}(C/|p(t)|_{\max}) \tag{5.4}$$

and $|p(t)|_{\max}$ is the maximum amplitude of the signal.

The value of CL was varied as 0 to −49 dB (7 dB step), so that eight stimuli were applied for test signals. As indicated in Table 5.1, pitch, signal duration, and listening level were fixed. Subjects participated were 19 students (male and female of 20 years of age) who listened to three stimuli and judged dissimilarity. The number of combinations of this experiment was ${}_8C_3 = 56$ triads. The dissimilarity matrix was

Table 5.1 Conditions of two experiments

Condition	Experiment 1	Experiment 2
(1) *Conditions fixed*		
Note (Pitch)	A4 (220 Hz) by use of 3rd string and 2nd fret	A4 (220 Hz) by use of 3rd string and 2nd fret
Listening level in L_{AE} (dB)	80	70
Signal duration (s)	4.0	1.5
(2) *Conditions varied*		
CL (dB) by Eq. (5.4)	Eight signals tested changing the cut-off level for 0–49 dB (7 dB step)	
Distortion type	–	Three different types: VINT, CRUNCH, and HARD
Drive level	–	Three levels due to the strength of distortion: 50, 70, 90 due to effectors type ME-30 (BOSE)

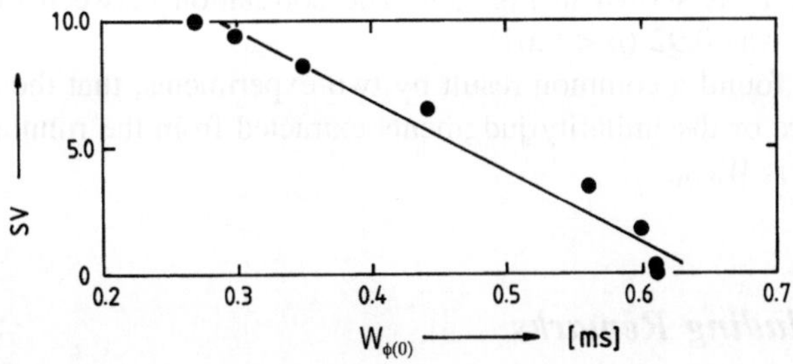

Fig. 5.8 Results for the SV obtained by regression analysis as a function of the mean value of $W_{\phi(0)}$ (Experiment 1)

made according to the judgments giving the number that, 2 for the most different pair, 1 for the neutral pair, and 0 for the most similar pair. After the analysis of multidimensional scaling, we obtained the SV.

We analyzed contributions to the SV of factors including the mean value of $W_{\phi(0)}$, the decay rate of SPL/s (dBA/s), and the mean value of ϕ_1 (pitch strength). It was found that the most significant factor contributing to the SV was the mean value of $W_{\phi(0)}$. Certain correlations between the mean value of $W_{\phi(0)}$ and other factors were found, but the mean value of $W_{\phi(0)}$ is considered as a representative factor. The SV as a function of the mean value of $W_{\phi(0)}$ is shown in Fig. 5.8. The correlation between the SV and the value of $W_{\phi(0)}$ was the most significant: 0.98 ($p < 0.01$).

Experiment 2

By the use of commercial effectors, this experiment is conducted finding the factor extracted from the running ACF contributing to the dissimilarity of sounds changing the strength of distortion. As indicated in Table 5.1, to produce nine

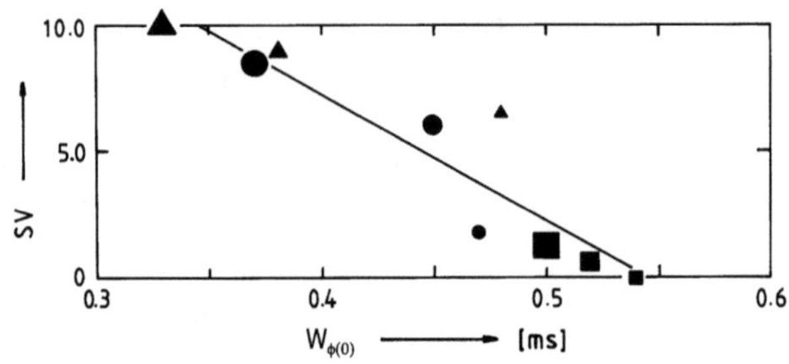

Fig. 5.9 Results for the SV obtained by regression analysis as a function of the mean value of $W_{\phi(0)}$ (Experiment 2)

stimuli with three kind of effect types (VINT, CHUTCH, and HARD) and three drive levels due to strength of distortion, i.e., 50, 70, and 90 by the effectors Type ME-30 (BOSE). Subjects participated were 20 students (male and female of 20 years of age). The method of experiment was same as above. Thus the number of combinations of this experiment was ${}_9C_3 = 84$ triads. Results achieved were similar to Experiment 1, as shown in Fig. 5.9. The correlation between the SV and the value of $W_{\phi(0)}$ was 0.92 ($p < 0.01$).

It has been found a common result by two experiments, that the most effective factor in timbre or dissimilarity judgments extracted from the running ACF of the source signal, is $W_{\phi(0)}$.

5.1.6 Concluding Remarks

It is obvious that above four primary percepts are all described by the factors extracted from the running ACF, as shown in Fig. 5.10. However, it has been difficult by applying any parameter from the analyses of spectrum of the sound signal. It is worth noticing that the single factor $W_{\phi(0)}$ is deeply related to a global frequency component of the source signal. It is noteworthy that Ohgushi (1980) showed that the lowest and highest frequency components govern primarily timbre. Remarkably, these two components may be replaced by the single factor $W_{\phi(0)}$ only.

5.2 Spatial Percepts in Relation to the Spatial Factors of the Sound Field

We shall show that spatial percepts are described by the spatial factors associated with the right cerebral hemisphere.

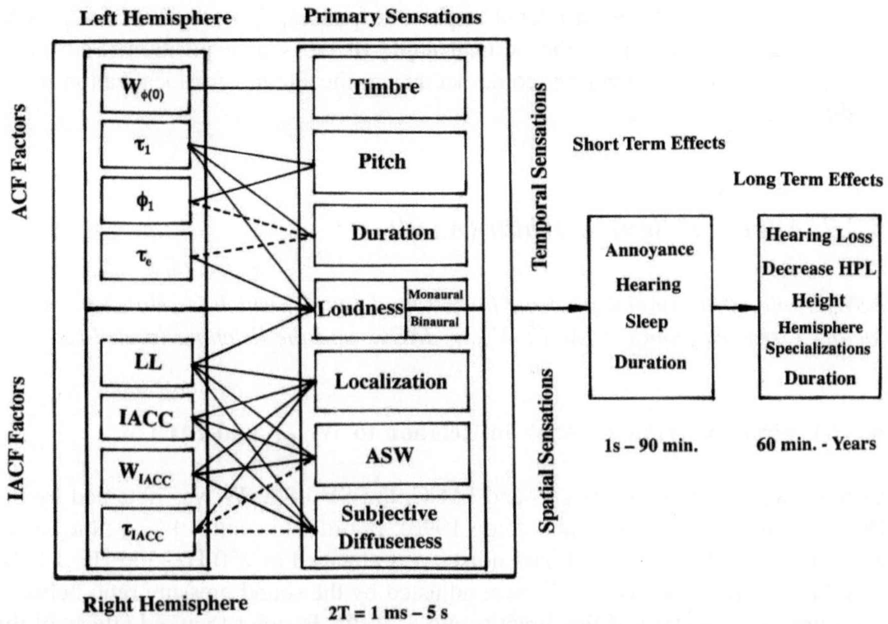

Fig. 5.10 Summarization of temporal and spatial sensations (percepts) in relation to the temporal factors extracted from the ACF and the spatial factors extracted from IACF, respectively

5.2.1 Localization of a Sound Source in the Horizontal and Median Plane

Localization of sound source is the most basic percept in spatial sensations. Let us now discuss localization in relation to the spatial factors, which can be extracted from the interaural cross-correlation function (IACF).

Localization of a sound source in the horizontal plane may be described essentially by the binaural and spatial factors, such that

$$L_{\text{horizontal plane}} = f[\Phi_{ll}(0),\ \Phi_{rr}(0),\ \tau_{IACC},\ \text{IACC},\ W_{IACC}], \tag{5.5}$$

where $\Phi_{ll}(0)$ and $\Phi_{rr}(0)$ are the sound energies arriving at two ear entrances, τ_{IACC}, IACC, and W_{IACC} are defined in Fig. 2.10. A movement of a sound source on the stage may be described by these spatial factors as a function of time.

As discussed below, significant spatial factors are IACC and W_{IACC} that influence on spatial percepts, such as subjective diffuseness and ASW of the sound field, respectively; and therefore sharp localizations may perceive with a large value of IACC and a narrow value of W_{IACC}.

Since the HRTF for a sound arriving at the both ears are similar in the median plane, such that $\Phi_{ll}(0) \approx \Phi_{rr}(0)$, $\tau_{IACC} \approx 0$, IACC ≈ 1.0 and W_{IACC} are all invariants for any position in this plane. However, it has been shown that the factors extracted from ACF of head-related impulse responses, τ_1, ϕ_1, and τ_e are dramatically different according to the vertical angle of the source position (Sato et al. 2001). These factors, therefore, could act as significant cues for localization in the median plane.

5.2.2 Apparent Source Width (ASW)

ASW is one of the spatial percepts for a sound source, which is related to spatial factors extracted from the IACF: W_{IACC}, IACC, and the listening level, LL.

5.2.2.1 First Experiment: ASW in Relation to W_{IACC} and IACC

Controlling the values of W_{IACC} and IACC, the SV of ASW was obtained by the PCT with ten subjects (Sato and Ando 1996). In order to control W_{IACC}, the center frequency of 1/3-octave band-pass noises was changed as 250 Hz, 500 Hz, 1 kHz, and 2 kHz. The values of IACC were adjusted by the sound pressure ratio between reflections ($\xi = \pm 54°$) and the direct sound ($\xi = 0°$). In order to avoid effects of the listening level on ASW (Keet 1968), in this investigation, the total SPL at the ear canal entrances of all sound fields was kept constant at a peak of 75 dBA. Listeners judged which of the two sound sources they perceived to be wider.

Results of the ANOVA for the SV S(ASW) indicate that both of the factors, IACC and W_{IACC}, are significant ($p < 0.01$) and contribute to S(ASW) independently, so that it yields

$$\begin{aligned} S_R(\mathrm{ASW}) &= f(\mathrm{IACC}) + f(W_{\mathrm{IACC}}) \\ &\approx \alpha(\mathrm{IACC})^{3/2} + \beta(W_{\mathrm{IACC}})^{1/2}, \end{aligned} \tag{5.6}$$

where coefficients $\alpha \approx -1.64$ and $\beta \approx 2.44$ are obtained by regressions of the SV with ten subjects as shown in Fig. 5.11. This holds under the conditions of a constant LL and $\tau_{IACC} = 0$. Obviously, as shown in Fig. 5.12, calculated SV by Eq. (5.6) and measured SV are in good agreement ($r = 0.97$, $p < 0.01$).

5.2.2.2 Second Experiment: ASW in Relation to W_{IACC} and LL

Controlling the values of W_{IACC} and LL, the SV of ASW was obtained by the PCT with five subjects (Sato and Ando 2002). In order to control W_{IACC}, we applied here the complex noise with different frequency components similar to above. To find the effects of LL on ASW, the SPL at the listener's head position was changed from 70 to 75 dB. The values of the IACC of all sound fields were fixed to 0.90 ± 0.01 by

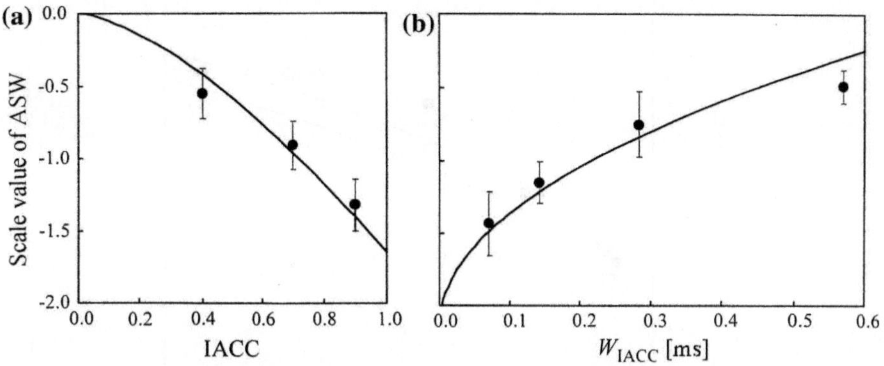

Fig. 5.11 SV of ASW for the 1/3 octave band-pass noises with 95 % reliability as a function of (**a**) the IACC and (**b**) the W_{IACC}

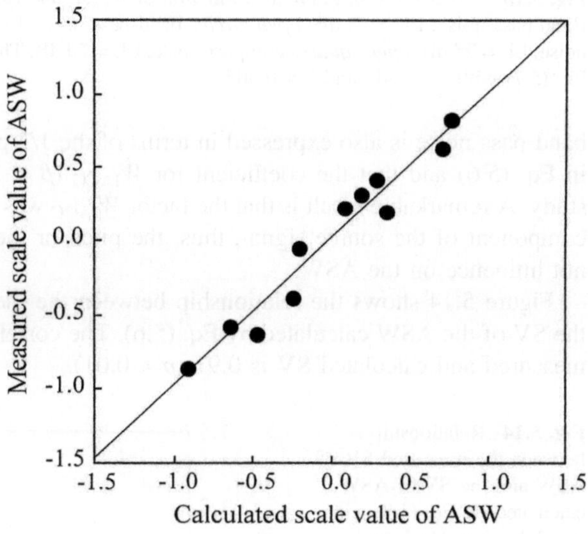

Fig. 5.12 Relationship between the measured SV of ASW and the SV of ASW calculated by Eq. (5.6) with α = 1.64 and β = 2.44. Correlation coefficient r = 0.97 ($p < 0.01$)

controlling the sound pressure ratio of the reflections relative to the level of the direct sound.

Results of the ANOVA for SV of the ASW revealed that the explanatory factors W_{IACC} and LL are significant ($p < 0.01$) (Fig. 5.13). The interaction between W_{IACC} and LL is insignificant, so that we obtain

$$\begin{aligned} S_R(\text{ASW}) = S_R &= f(W_{IACC}) + f(\text{LL}) \\ &\approx a(W_{IACC})^{1/2} + b(\text{LL})^{3/2}, \end{aligned} \tag{5.7}$$

where coefficients a = 2.40 and b = 0.005, coefficients were obtained by multiple regression analyses. It is noteworthy that the SV of ASW for the 1/3-octave

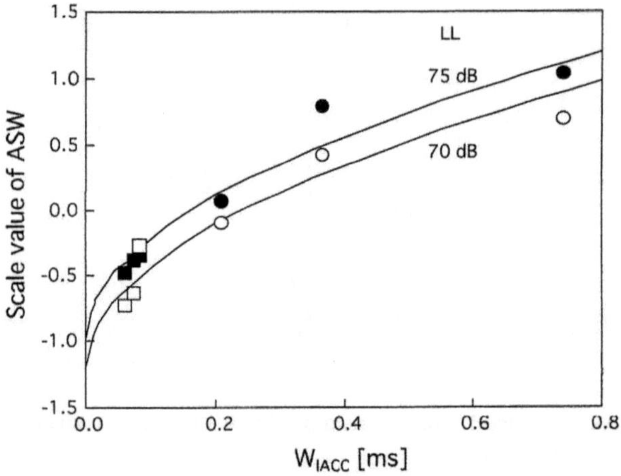

Fig. 5.13 Average SV of ASW as a function of W_{IACC} and as a parameter of LL. *Filled circles* Band-pass noise; LL = 75 dB; *open circles* band-pass noise; LL = 70 dB; *filled squares* complex noise; LL = 75 dB; *open squares* complex noise; LL = 70 dB. The regression curve is expressed by Eq. (5.7) with a = 2.40, and b = 0.005

band-pass noise is also expressed in terms of the 1/2 power of W_{IACC} as expressed in Eq. (5.6) and that the coefficient for W_{IACC} ($\beta \approx 2.44$) is close to that of this study. A remarkable result is that the factor W_{IACC} was determined by the frequency component of the source signal; thus, the pitch or the fundamental frequency did not influence on the ASW.

Figure 5.14 shows the relationship between the measured SV of the ASW and the SV of the ASW calculated by Eq. (5.6). The correlation coefficient between the measured and calculated SV is 0.97 ($p < 0.01$).

Fig. 5.14 Relationship between the measured SV of ASW and the SV of ASW calculated by Eq. (5.8) with a = 2.40, b = 0.005, and c = −1.60. The correlation coefficient, $r = 0.97$ ($p < 0.01$)

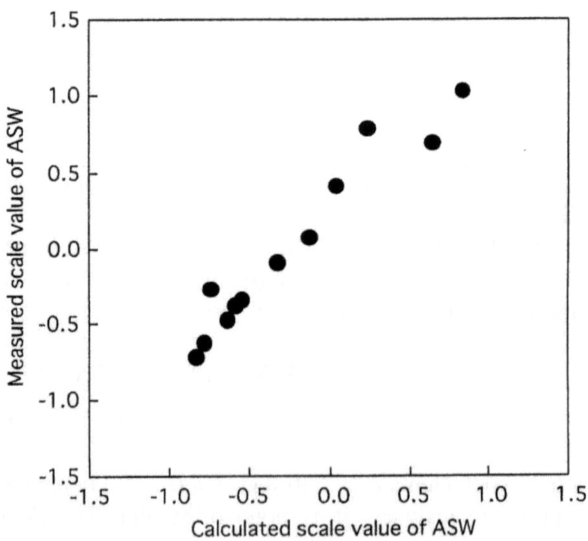

5.2.2.3 ASW in Relation to All of the Three Factors, W_{IACC}, LL, and W_{IACC}

It is interesting that the weighting coefficients of $(W_{IACC})^{1/2}$ in two different experimental results are apparently similar around 2.40. If we combine two Eqs. (5.6) and (5.7) into one, a single formula yields

$$S(\mathrm{ASW}) = S_R = f(W_{IACC}) + f(\mathrm{LL}) + f(\mathrm{IACC}) \approx a(W_{IACC})^{1/2} + b(\mathrm{LL})^{3/2} + c(\mathrm{IACC})^{3/2}, \quad (5.8)$$

where coefficients are: $a \approx 2.40$, $b \approx 0.005$ and $c \approx -1.60$.

At the design stage of an opera house, for example, the SV of ASW may be calculated after getting the spatial factors extracted from IACF at each seating position by the use of architectural scheme.

5.2.3 Subjective Diffuseness

Subjective Diffuseness that deeply related to spatial impression and envelopment is representative percept of spatial sensation of the sound field in an opera house.

In order to obtain the SV of subjective diffuseness, the PCT with bandpass Gaussian noise, varying the horizontal angle of two symmetric reflections, has been conducted (Ando and Kurihara 1986; Singh et al. 1994). Listeners judged which of

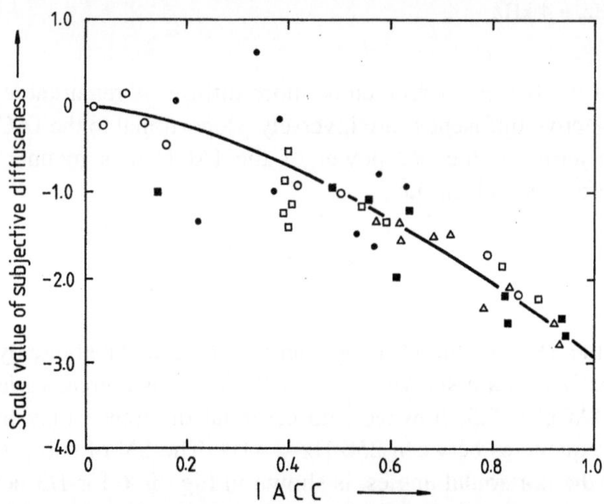

Fig. 5.15 SV of subjective diffuseness as a function of the IACC (calculated). *Different symbols* indicate different frequencies of the 1/3 octave bandpass noise: *open triangles* 250 Hz, *open circles* 500 Hz, *open squares* 1 kHz, *filled circles* 2 kHz, *filled squares* 4 kHz. *Line* Regression line by Eq. (5.9)

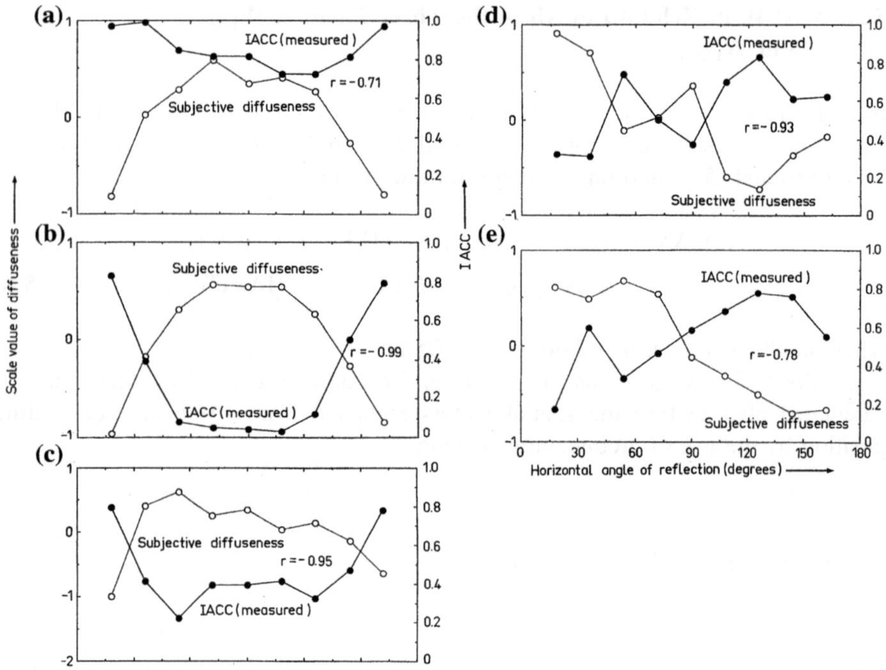

Fig. 5.16 SV of subjective diffuseness and the IACC as a function of the horizontal angle of incidence to a listener, with 1/3 octave band noise of center frequencies. **a** 250 Hz. **b** 500 Hz. **c** 1 kHz. **d** 2 kHz. **e** 4 kHz

the two sound fields were perceived as more diffuse. A remarkable finding is that the SV of subjective diffuseness are inversely proportional to the IACC, and may be formulated in terms of the 3/2 power of the IACC in a manner similar to the subjective preference values, i.e.,

$$S \approx -\alpha(\mathrm{IACC})^{\beta}, \tag{5.9}$$

where $\alpha = 2.9$, $\beta = 3/2$.

Results of the SV from the PCT together and the calculated one by Eq. (5.9), as a function of the IACC, are shown in Fig. 5.15. There is a great variation of data in the range of IACC < 0.5; however, no essential difference may be found in the results with frequencies between 250 Hz–4 kHz. The SV of subjective diffuseness depending on the horizontal angles, is shown in Fig. 5.16 for 1/3 octave band-pass noise with the center frequencies of 250 Hz, 500 Hz, 1 kHz, 2 kHz, and 4 kHz. Clearly, the most effective horizontal angles of reflections depend on the frequency range. As shown in Fig. 5.17, for example, these are about ±90° for the 500 Hz range and the frequency range below 500 Hz, around ±55° for the 1 kHz range

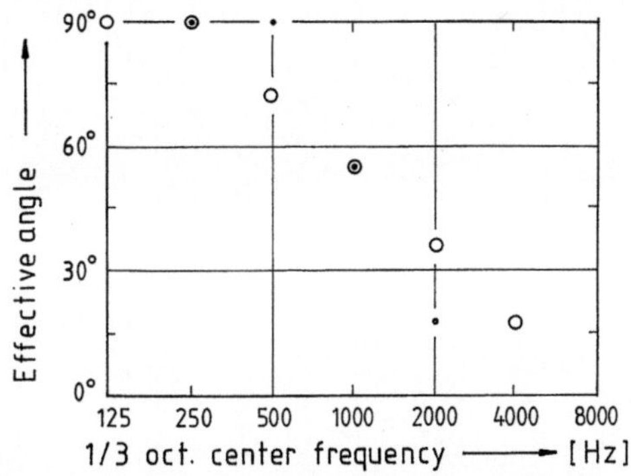

Fig. 5.17 The most effective horizontal angles to a listener, making decrease of the IACC for each frequency band, obtaining the maximum SV of subjective diffuseness. *Open circles* Angles obtained by the calculated IACC; *filled circles* angles obtained by the observed IACC

(that is the most important angle for the music), and smaller than ±20° for the 2 kHz and 4 kHz ranges.

So far, we discussed temporal primary percepts and spatial primary percepts in relation to temporal factors extracted from ACF of sound signals and spatial factors from IACF, respectively, as summarized in Fig. 5.10.

Chapter 6
Theory of Subjective Preference of the Sound Field

Subjective preference judgment is the most primitive response and the overall response. It is regarded in living creatures as one that entails judgments that steer an organism in the direction of maintaining life, so as to enhance its prospects for survival. Also, it is interesting that the preference judgment is deeply related to the aesthetic issue. The method of obtaining the linear scale value of stimuli by applying the law of comparative judgment is well known (Thurston 1927; Gulikusen 1956; Torgerson 1958). The PCT is the simplest method, so that any person may participate and results may be integrated and utilized in a wide range of applications. From results of subjective preference studies in relation to temporal factors and spatial factors of the sound field, the theory of subjective preference is established. Based on the theory, an example of calculating subjective preference at each listener's position is demonstrated.

6.1 Sound Fields with a Single Reflection and Multiple Reflections

6.1.1 Preferred Delay Time of a Single Reflection

To begin with investigations for the simplest sound fields with a single reflection, we can obtain basic knowledges toward establishing a theory of subjective preference of the sound field with early multiple reflections and subsequent reverberation of enclosures (Ando 1985, 1998).

The sound field consists of the direct sound $\xi_1 = 0°$ $(\eta_1 = 0°)$ and a single reflection from a fixed direction $\xi_1 = 36°$ $(\eta_1 = 90°)$. These angles were selected, since they are typical in a concert hall. The delay time Δt_1 of the single reflection after arriving of the direct sound was adjusted in the range of 6–256 ms. Paired comparison tests (PCT) were performed with doctorate students and assistants at the Third Physics Institute in Gottingen with two different music motifs A and B (Table 6.1). The score is obtained here by accumulating score giving +1 and −1 corresponding to positive and negative judgments, respectively, and the total score

Y. Ando, *Opera House Acoustics Based on Subjective Preference Theory*,
Mathematics for Industry 12, DOI 10.1007/978-4-431-55423-3_6

Table 6.1 Music and speech source signals and their minimum effective duration of the running ACF, $(\tau_e)_{min}$

Sound source	Title	Composer or writer	$(\tau_e)_{min}$ (ms)[a]
Music motif A	Royal Pavane	Orlando Gibbons	125
Music motif B	Sinfonietta, opus 48; IV movement	Malcolm Arnold	40
Speech S	Poem read by a female	Doppo Kunikida	10

[a] The value of $(\tau_e)_{min}$ is the minimum value extracted from the running ACF, $2T = 2$ s, with a running interval of 100 ms

is divided by $S(F - 1)$ to get the normalized score, where S is the number of subjects and F is the number of sound fields. The normalized scores and the percentage of preference of the sound fields as a function of the delay are shown in Fig. 6.1a, b (Ando 1977; Ando and Kageyama 1977; Ando and Morioka 1981; Kang and Ando 1985).

Obviously, the most preferred delay time, with the maximum score, differs greatly between the two motifs and the speech signal. When the amplitude of reflection is $A_1 = 1$, the most preferred delays are around 130 ms for music motif A, 35 ms for music motif B (Fig. 6.1a), and 16 ms for the speech (Fig. 6.1b). It was found that these correspond well to the effective duration of the ACF, $(\tau_e)_{min}$, of

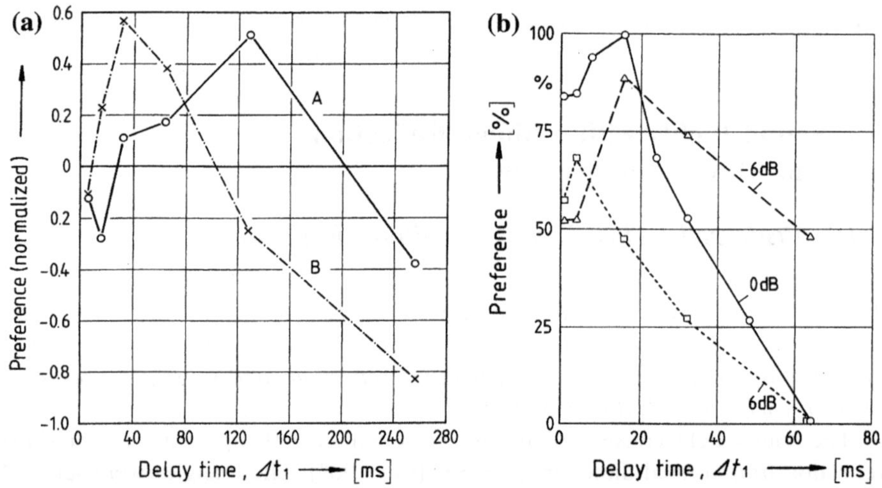

Fig. 6.1 Results of subjective preference for different sound sources as a function of the delay time of single reflection obtained by the PCT giving simply +1 and −1, corresponding to the positive and negative judgment, respectively. The normalized score is obtained by the factor $S(F - 1)$, S being the number of sound field and F is the number of subjects (6 sound fields and 13 subjects). **a** Normalized preference score for music motif A ($\tau_e \approx 127$ ms) and music motif B ($\tau_e \approx 35$ ms) (Ando 1977). **b** Percentile preference for a continuous speech signal ($\tau_e \approx 12$ ms). It is worth noticing that the most preferred delay time of the reflection is related to approximately the value of τ_e, when $A_1 = 1.0$ (Ando and Kageyama 1977)

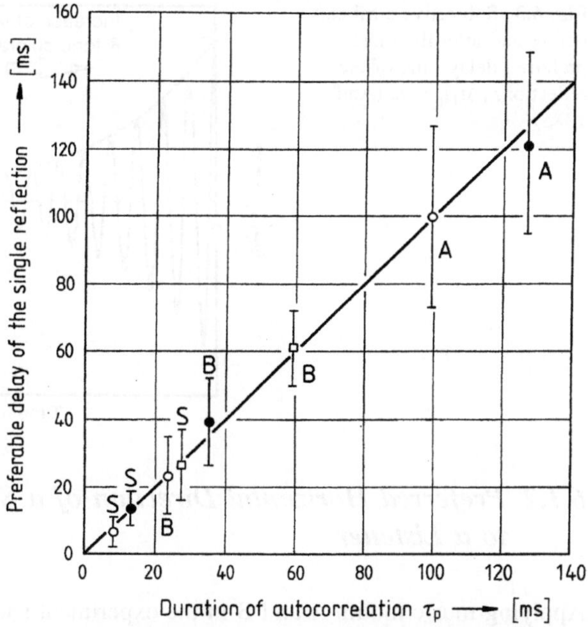

Fig. 6.2 The most preferred delay time of the single reflection as a function of the duration of the autocorrelation function of source signals, $\tau_p = [\Delta t_1]_p$ is given by Eq. (6.7)

source signals of 125 ms (motif A), 40 ms (motif B) and 10 ms (Speech) as indicated in Table 6.1. After inspection, the preferred delay was found roughly at a certain duration of ACF, defined by τ_p, such that the envelope of ACF becomes $0.1A_1$. Thus, $\tau_p = (\tau_e)_{min}$ only when $A_1 = 1$. The data collected as a function of the duration τ_p are shown in Fig. 6.3, where data from a continuous speech signal are also plotted. When the envelope of the ACF is exponential, then it is expressed approximately by (Ando 1985) (Fig. 6.2).

$$\tau_p = [\Delta t_1]_p \approx (1 - \log_{10} A_1)(\tau_e)_{min} \quad (6.1)$$

It is worth noticing that the amplitude of reflection relative to that of the direct sound should be measured by the most accurate method, for example, the square root of ACF at the origin of the delay time, $[\Phi_p(0)]^{1/2}$.

Two reasons can be given for explaining why the preference decreases for the short delay range of reflection, $0 < \Delta t_1 < \tau_p = [\Delta t_1]_p$ (Fig. 6.3):

(1) Tone coloration effects occur because of the interference phenomenon in the coherent time region.
(2) The IACC increases when Δt_1 is near 0.

On the other hand, echo disturbance effects can be observed when Δt_1 is greater than τ_p.

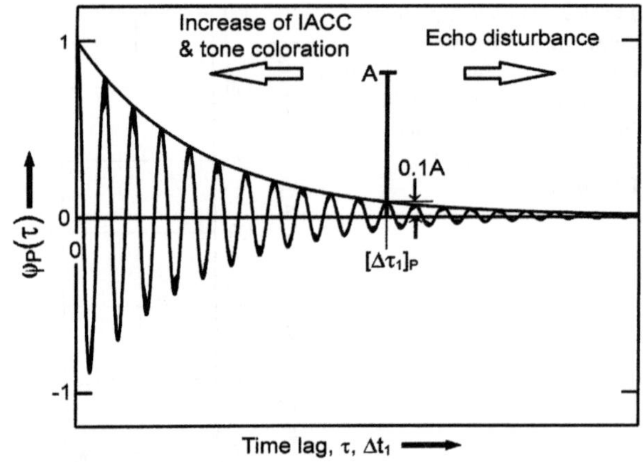

Fig. 6.3 Subjective attributes before and after the most preferred delay time of the reflection, $[\Delta t_1]_p = \tau_p$ (Ando 1985)

6.1.2 Preferred Horizontal Direction of a Single Reflection to a Listener

Applying music motifs A and B in the experiment showing the preferred direction of a single reflection, the delay time of the reflection was fixed at 32 ms. The direction was specified by loudspeakers located at

$$\xi_1 = 0° \ (\eta_1 = 27°), \text{ and } \xi_1 = 18°, 36°, \ldots, 90° \ (\eta_1 = 9°).$$

Results of the preference tests are shown in Fig. 6.4. No fundamental differences are observed between the curves of the two motifs in spite of the great difference in $(\tau_e)_{min}$. The preferred score increases roughly with decreasing IACC. The correlation coefficient between the score and the IACC is −0.8 ($p < 0.01$). The score with motif A at $\xi_1 = 90°$ drops to a negative value, indicating that the lateral reflections, coming only from around $\xi_1 = 90°$, are not always preferred. The figure shows that there is a preference for angles less than $\xi_1 = 90°$, and on average there may be an optimum range centered on about $\xi_1 = 55°$. Similar results can be seen in the data from speech signals (Ando and Kageyama 1977).

6.2 Sound Fields with Early Reflections and the Subsequent Reverberation

In order to examine the independence of the effects of the four physical factors on subjective preference judgments, two of them were varied simultaneously while the remaining two were held constant. The experiments were repeated thrice (Tests A, B, and C) obtaining four-dimensional continuity of the scale value.

Simulating the sound fields in opera house, a computer program provides the time delay of two early reflections ($n = 1, 2$) and the subsequent reverberation ($n > 2$) , relative to the direct sound. In order to represent the geometrical size of a similar room shape, the scale of dimension (SD) is introduced as follows:

$$\Delta t_1 = 22(\text{SD}),\ \Delta t_2 = 38(\text{SD}),\ \Delta t_3 = 47(\text{SD}) \quad (\text{ms}) \tag{6.2}$$

The reverberation signal with constant frequency characteristics was generated by the Schroeder reverberator (Schroeder 1962). To obtain a natural sound, the conditions of the simulation system were carefully selected.

Test A

In order to examine whether or not the temporal-monaural factors influence the scale values of subjective preference independently, PCTs of the 16 sound fields for each source signal were conducted for changes of (SD) and T_{sub} with 9–14 subjects (Ando et al. 1982). Both the factors are associated with the left hemisphere of the human brain, as mentioned in Chap. 4.

Test B

In order to determine independent influence of the spatial and right hemispheric factors, LL and IACC, PCTs of the 12 sound fields for each source signal were performed with 13–14 subjects (Ando and Morioka 1981).

Test C

For reconfirmation of the independence of the left and right hemispheric factors, T_{sub} and IACC, preference tests were conducted for 16 sound fields with 8 subjects (Ando et al. 1983).

According to results of analyses of variance for the scale values obtained by the law of comparative judgments from three tests, each factor influences the scale value of preference independently (Ando 1985).

Other Tests

In addition, subjective preference judgments performed in sound fields with multiple early reflections (Ando and Gottlob 1979), and with subsequent reverberation (Ando and Imamura 1979) confirmed that the factors (SD) and the IACC are independent of each other in the subjective preference judgments.

Thus, it may be concluded that the total scale value of subjective preference is determined by the law of superposition in the range of preferred conditions of the four factors. The consistency of the unit of the scale values obtained from the different preference tests has been discussed at length (Ando 1985).

6.3 Optimal Conditions Maximizing Subjective Preference

According to the above systematic investigations of simulating sound fields in an opera house with the aid of a computer and listening tests (PCT), the optimum design objectives and the linear scale value of subjective preference may be

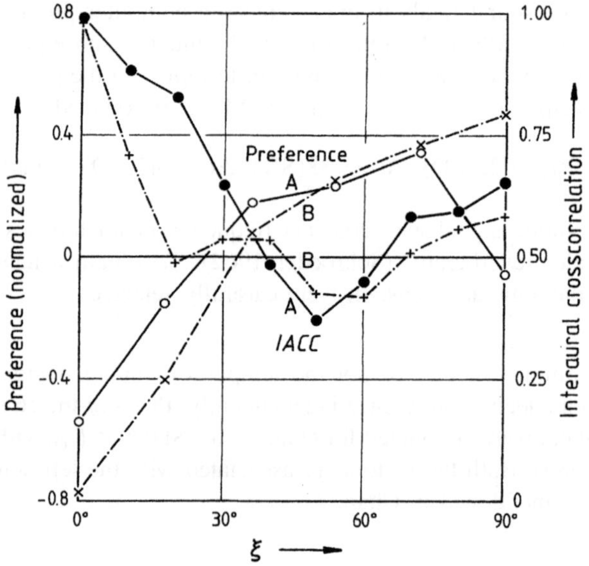

Fig. 6.4 Normalized preference score and the IACC for music motif A ($\tau_e \approx 127$ ms) and music motif B ($\tau_e \approx 35$ ms) as a function of the horizontal angle of incidence of the single reflection, ξ (Ando 1977). The negative strong correlation between subjective preference and IACC may be found. The most effective horizontal angle minimizing the IACC may be found commonly around $\xi \approx 55°$ regardless of sound source or the value of τ_e

described. The optimum design objectives can be described in terms of the subjectively preferred sound qualities, which are related to the temporal and spatial factors describing the sound signals arriving at the two ears. They clearly lead to comprehensive criteria consisting of the temporal and spatial factors associated with the left and the right human cerebral hemispheres, respectively. For achieving the optimal design of opera houses, we shall summarize optimal conditions of the four orthogonal factors below (Ando 1983, 1985, 1998).

6.3.1 Listening Level (LL)

The listening level is, of course, the primary criterion for listening to vocal and orchestra music in the sound field of opera houses. The preferred listening level depends on the music and the particular passage being performed. For example, the gross preferred levels obtained by 16 subjects are in peak ranges of 77–79 dBA for two different musical sound sources, i.e., music motif A (Royal Pavane by Gibbons) with a slow tempo, and 79–80 dBA for music motif B (Sinfonietta by Arnold) with a fast tempo.

6.3.2 Early Reflections After the Direct Sound (Δt_1)

An approximate relationship for the most preferred delay time $[\Delta t_1]_p$ has been developed in terms of the envelope of autocorrelation function of source signals and the total amplitude of reflections, A (Ando 1985). Generally, it is expressed by

$$\left|\phi_p(\tau)\right|_{\text{envelope}} \sim kA^c, \text{ at } \tau = [\Delta t_1]_p \tag{6.3}$$

where k and c are constants, which depend on the subjective attributes (Table 6.2 in Ando 1998). When the envelope of ACF is exponential as indicated for most music and speech signals, then

$$[\Delta t_1]_p \approx (\log_{10} 1/k - c \log_{10} A)(\tau_e)_{\min} \tag{6.4}$$

where the value of $(\tau_e)_{\min}$ is minimum effective duration of running ACF and the total pressure amplitude of reflection is the sum of each normalized amplitude of reflections by that of the direct sound A_0, A_n ($n > 1$), such that

$$A = \left[A_1^2 + A_2^2 + A_3^2 + \cdots\right]^{1/2} \tag{6.5}$$

The relationship of Eq. (6.1) for a single reflection may be obtained by putting $A = A_1$, $k = 0.1$ and $c = 1$ so that

$$[\Delta t_1]_p \approx (1 - \log_{10} A_1)(\tau_e) \tag{6.6}$$

It has been accurately expressed that the value of τ_e in Eq. (6.6) is replaced by the minimum value of τ_e of the running ACF (Ando et al. 1989; Mouri et al. 2000), so that

$$[\Delta t_1]_p \approx (1 - \log_{10} A_1)(\tau_e)_{\min} = \tau_p \tag{6.7}$$

The value of $(\tau_e)_{\min}$ is usually observed in the music at the most active part containing the most "artistic" information like a "vibrato", a "quick" in the music flow and/or a very sharp sound signal like attack. The echo disturbance, therefore, may perceive at the signal piece occurring $(\tau_e)_{\min}$.

Even for a long music composition, the music flow might be divided into such a short piece that the most minimum part of $(\tau_e)_{\min}$ of the running ACF is in the whole music, which determines the preferred temporal conditions. It may be taken into consideration for the choice of music program to be performed in a given opera house. A method of controlling the minimum value $(\tau_e)_{\min}$, which determines the preferred temporal conditions for vocal music has been discussed (Sect. 9.3; Kato and Ando 2002; Kato et al. 2004). If the vibrato is introduced during singing, then it effectively decreases $(\tau_e)_{\min}$ blending the short reverberation time of the usual opera house.

6.3.3 Subsequent Reverberation Time After the Early Reflections (T_{sub})

It has been examined that the most preferred conditions of frequency response to the reverberation time is just flat (Ando et al. 1989). The preferred subsequent reverberation time is expressed approximately by

$$[T_{\text{sub}}]_{\text{p}} \approx 23(\tau_{\text{e}})_{\text{min}} \tag{6.8}$$

The values A given by Eq. (6.5) tested were 1.1 and 4.1, which cover the usual conditions of sound fields in a room.

Recommended reverberation times for several sound sources are shown in Fig. 6.5. A lecture and conference room must be designed for speech. And for an opera house for two greatly different sources, i.e., vocal and orchestra music could be adjusted at short reverberation time and at relatively long value, respectively (Sect. 10.5).

6.3.4 Magnitude of the Interaural Cross-Correlation Function (IACC)

All available data over 500 listeners tested indicated a negative correlation between the magnitude of the interaural cross-correlation function (IACC) and the subjective preference, i.e., dissimilarity of signals arriving at the two ears is preferred (Schroeder et al. 1974). This holds only under the condition that the maximum value of the IACC is maintained at the origin of the time delay keeping balance of

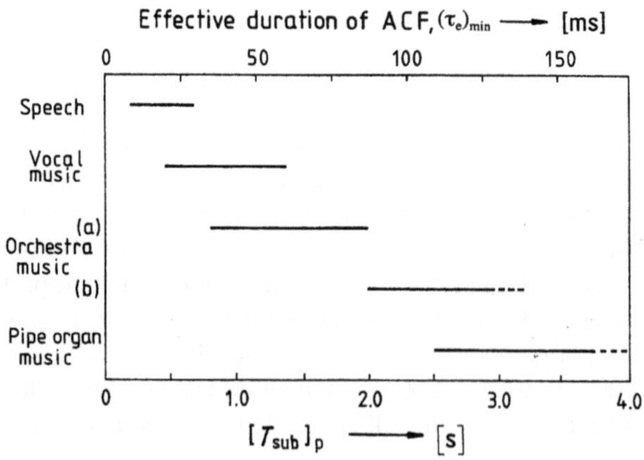

Fig. 6.5 Reverberation time recommended for each sound source that is related to the value of τ_e given by Eq. (6.8)

sound field for two ears. If not, then an image shift of the source may occur. To obtain a small magnitude of IACC in the most effective manner as shown in Fig. 5.17, the directions from which the early reflections arrive at the listener should be kept within a certain range of angles from the median plane for usual sound sources consisting of main frequency range centered on 1.0 kHz as $\pm(55° \pm 20°)$. It is obvious that the sound arriving from the median plane ±0° makes the IACC greater. Sound arriving from ±90° in the horizontal plane is not always advantageous, because the similar "detour" paths around the head to both ears cannot decrease the IACC effectively, particularly for frequency ranges higher than 500 Hz. For example, the most effective angles for frequency ranges of 1 and 2 kHz are centered about ±55° and ±36°, respectively.

In order to realize these conditions simultaneously, a geometrical uneven surface has been proposed (Ando and Sakamoto 1988).

6.4 Theory of Subjective Preference for the Sound Field

Let us now see these results with the temporal factors and the spatial ones associated with the left and the right human cerebral hemispheres, respectively (Table 4.1), put into a theory of subjective preference.

Since the numbers of orthogonal acoustic factors, which are included in the sound signals at both ears, are limited to four as mentioned in Sect. 3.2 (Ando 1983, 1985, 1998), the scale value of any one-dimensional subjective response may be expressed as

$$\begin{aligned} S &= g(x_1, x_2, \ldots, x_I) \\ &\equiv g_L + g_R \end{aligned} \tag{6.9}$$

where L and R signify the left and right specializations of the human cerebral hemisphere. In this study, the linear scale value of preference obtained by the law of comparative judgment is described. It has been verified by a series of experiments that four objective factors act independently of the scale value when changing two of four factors simultaneously. Results indicate that the units of scale values are considered to be almost constant, so that we may add scale values to obtain the total scale value (Ando 1983),

$$\begin{aligned} S &\equiv g_L + g_R = g(x_1) + g(x_2) + g(x_3) + g(x_4) \\ &= S_1 + S_2 + S_3 + S_3 \end{aligned} \tag{6.10}$$

where $g_L = S_2 + S_3$ and $g_R = S_1 + S_4$ and S_i, $i = 1, 2, 3, 4$ is the scale value obtained relative to each objective parameter. Equation (6.10) indicates a four-dimensional continuity. The scale value is relative and addition and subtract operations only may be allowed.

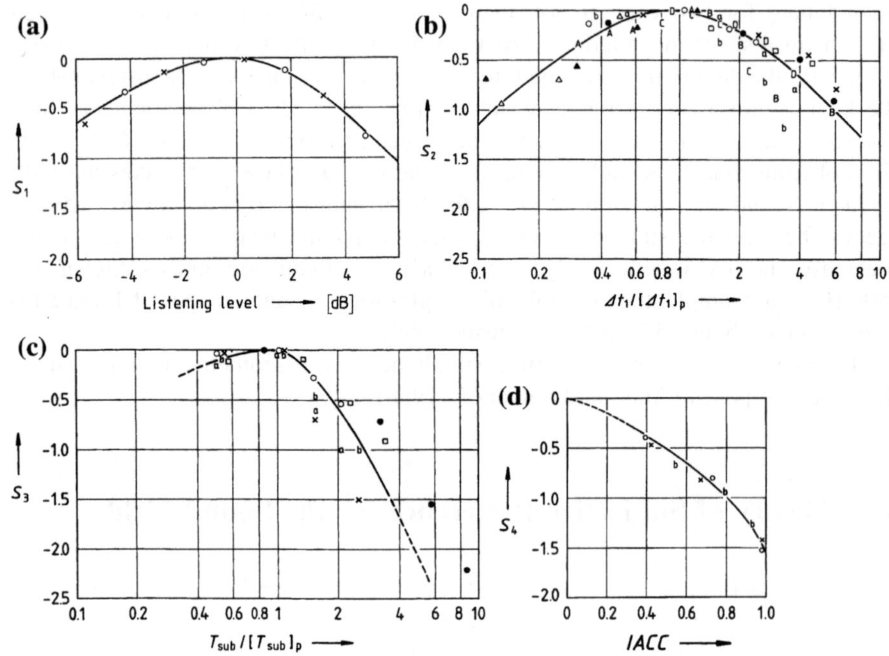

Fig. 6.6 The scale value of subjective preference as a function of each of four orthogonal factors of the sound field. The maximum scale value is adjusted to zero at the most preferred condition of each factor. **a** Scale value S_1 as a function of the LL. **b** Scale value S_2 as a function of the Δt_1 normalized by $[\Delta t_1]_p$ calculated by Eq. (6.7). **c** Scale value S_3 as a function of the T_{sub} normalized by $[T_{sub}]_p$ calculated by Eq. (6.8). **d** Scale value S_3 as a function of the IACC

The dependence of the scale values on each of four orthogonal factors is shown graphically in Fig. 6.6a–d. From the nature of the scale value, it is convenient to put a zero value at the most preferred conditions, as shown in these figures. These results of the scale value of subjective preference obtained from the different test series, using different music programs, yield the following common formula as indicated by solid lines in Fig. 6.6a–d:

$$S_i \approx -\alpha_i |x_i|^{3/2}, \quad i = 1, 2, 3, 4 \tag{6.11}$$

where x_i is, and the values of α_i are weighting coefficients as listed in Table 6.2. If α_i is close to zero, then a lesser contribution of the factor x_i on subjective preference is signified.

The factor x_1 is given by the sound pressure level difference, measured by the A-weighted network, so that

$$x_1 = 20\log P - 20\log [P]_p \tag{6.12}$$

Table 6.2 Four orthogonal factors or design criteria of the sound field and its weighting coefficients α_i in Eq. (6.3), which were obtained by a series of experiments (the paired comparison test) on subjective preference with a number of subjects (Ando 1983, 1998)

I	Four orthogonal factors, x_i (i = 1, 2, 3, 4)	α_i	
		$X_i > 0$	$x_i < 0$
1	$20 \log P - 20 \log [P]_p$ (dB)	0.07	0.04
2	$\log (\Delta t_1/[\Delta t_1]_p)$	1.42	1.11
3	$\log (T_{sub}/[T_{sub}]_p)$	$0.45 + 0.75A$	$2.36 - 0.42A$
4	IACC	1.45	–

P and $[P]_p$ being the sound pressure at a specific seat and the most preferred sound pressure that may be assumed at a particular seat position in the room under investigation;

$$x_2 = \log\left(\Delta t_1/[\Delta t_1]_p\right) \tag{6.13}$$

$$x_3 = \log\left(T_{sub}/[T_{sub}]_p\right) \tag{6.14}$$

$$x_4 = \text{IACC} \tag{6.15}$$

Thus, the scale values of preference have been formulated approximately in terms of the 3/2 power of the normalized objective parameters, expressed in the logarithm for the parameters, x_1, x_2 and x_3. The remarkable fact is that the spatial binaural parameter x_4 is expressed in terms of the 3/2 powers of its "real values," indicating a greater contribution than those of other parameters. Thus, the scale values are not greatly changed in the neighborhood of the most preferred conditions, but decrease rapidly outside of this range. Since the experiments were conducted to find the optimal conditions, this theory holds in the range of preferred conditions tested for the four factors.

This theory has been well based on neural activities in the auditory-brain system that is deeply related to subjective preference of the sound field (Chap. 4). It has been clearly shown the two temporal factors the initial time delay gap between the direct sound and the first reflection Δt_1 and the subsequent reverberation time T_{sub} are associated with the left cerebral hemisphere, and the two spatial factors the magnitude of IACF (IACC) and the listening level LL are associated with the right.

Applying the theory of subjective preference, we demonstrate the quality of the sound field at each seating position in a typical and simple room shape similar to that of Symphony Hall in Boston. Suppose that a single source is located at the center, 1.2 m above the stage floor. Receiving points at a height of 1.1 m above the floor level correspond to the ear positions. Reflections with their amplitudes, delay times, and directions of arrival at the listeners are taken into account using the image method.

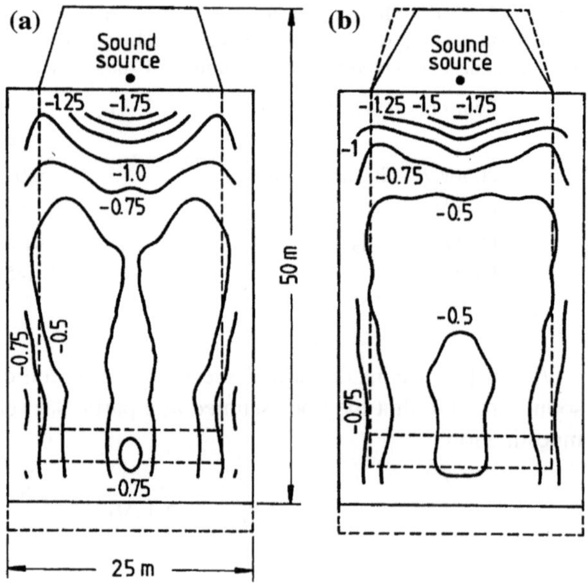

Fig. 6.7 Examples of calculating the total scale value after obtaining the four orthogonal factors at each seating position. **a** An original room shape. **b** Stage enclosure adjusted

Contour lines of the total scale value of preference calculated for music motif B are shown in Fig. 6.7. This figure shows effects of the reflection from the side reflections on the stage. The sidewalls on the stage may produce decreasing values of IACC for the audience area. Thus, the preference value at each seat is increased, as shown in Fig. 6.7b in comparison with the original one (Fig. 6.7a). In this calculation, the reverberation time is assumed to be 1.8 s throughout the hall and the most preferred listening level, $[LL]_p = 20 \log [P]_p$ in Eq. (6.12), is assumed for a point on the center line 20 m from the source position.

Chapter 7
Examination of Subjective Preference Theory in an Existing Opera House

We examine whether or not the subjective preference theory holds for the sound field in an existing opera house. Here, Teatro Comunale in Ferrara is selected for the purpose of this study. Paired-comparison tests (PCT) were conducted to obtain scale value of subjective preference, switching loudspeakers (or source locations) of the music on the stage and those in the orchestra pit. Subjects were asked which of the two sound fields they preferred to listen. The orthogonal acoustical factors of the sound field at each listening position were obtained from binaural impulse responses and the interaural cross-correlation function (IACF) measured at each listening position. The relationship between the scale value of subjective preference and orthogonal acoustical factors was obtained by using the factor analysis. Results show that the total scores obtained from factor analysis and measured scale values are in good agreement.

7.1 Measurement of Orthogonal Factors of the Sound Field at Each Seat

7.1.1 Procedure

Acoustical measurements were conducted to obtain the orthogonal factors at each listening position based on the four conditions, arranging sound sources located in the pit and on the stage as listed in Table 7.1 and shown in Fig. 7.1 (Sato et al. 2002). Acoustic conditions were the same as those applied for the subjective preference judgments. Definitions and calculation procedures of the acoustical factors are described in Sect. 3.2.

To obtain the spatial factors extracted from the IACF, the musical signals used in the subjective tests were reproduced using the same loudspeaker configurations as in the subjective tests. Two channel signals were recorded at each listening position into a digital audio tape (DAT) through two condenser microphones at the entrance of two ears of a dummy head (Appendix), which is facing the center of the stage. In the measurements in the boxes, the receiver's head was located at the opening plane

Y. Ando, *Opera House Acoustics Based on Subjective Preference Theory*,
Mathematics for Industry 12, DOI 10.1007/978-4-431-55423-3_7

Table 7.1 Four conditions of the loudspeaker positions in the paired-comparison tests

	Condition 1	Condition 2	Condition 3	Condition 4
Stage	Front	Rear	Front	Rear
Orchestra pit	Front	Front	Rear	Rear

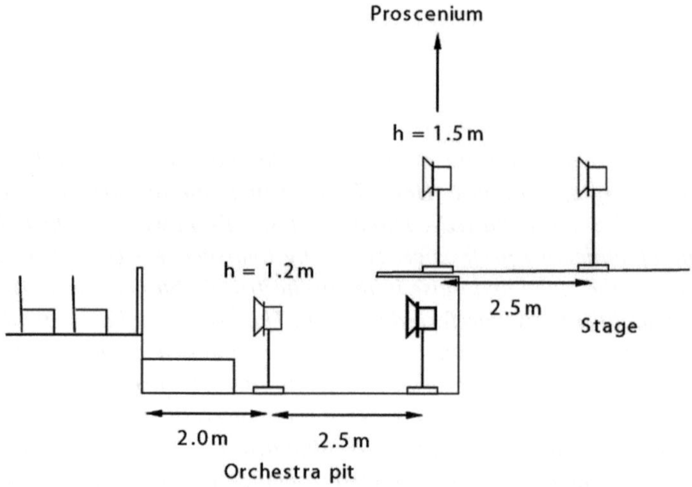

Fig. 7.1 Locations of the loudspeakers on the stage and in the orchestra pit

of the box. The IACFs were obtained after passing through an A-weighting network. Four orthogonal factors, namely biaural listening level (LL) and IACC, in addition to W_{IACC} and interaural time delay (τ_{IACC}) were extracted from the IACF.

To obtain the temporal factors which may be obtained from the impulse response, a log sine sweep with a duration 15 s was applied. For each position where the subjects were seated, Δt_1 and T_{sub} were obtained from the impulse response for each sound source.

7.1.2 Measurement Results

(A) Spatial Factors
The maximum LL was observed under condition 1 at seat 4 in the stall (Fig. 7.2). The range of the values of IACC was between 0.09 (Condition 2 at seat 5 in the stall) and 0.65 (Condition 3 at seat 1 in the stall). The τ_{IACC} values of almost all listening positions were less than 0.10 ms because the receiver faced the sources during the measurement. The W_{IACC} values were centered on 0.12 ms for all the four conditions.

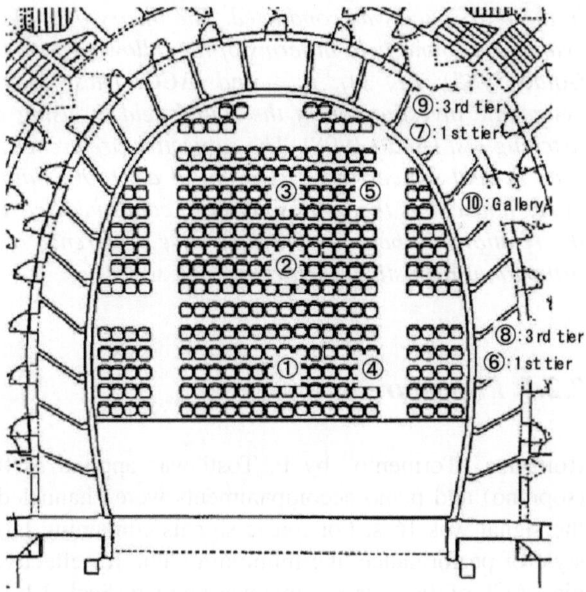

Fig. 7.2 Plan of Teatro Comunale in Ferrara. The numbers in the *circle* indicate seating positions of subjects and measuring the physical factors of the sound fields

(B) Temporal Factors
Values of Δt_1 in the stall for the source in the pit are almost similar to around 10 ms for the frontal source, and 3 ms for the rear source. These reflections may come from the rear wall of the pit. Values of Δt_1 in the stall for the source on the stage are longer than those for the source in the pit. The reflection may arrive from the sidewall of the audience area, when the sources located on the stage. In the boxes, the first reflections come from the walls inside the box. In the gallery ⑩, the first reflections for all four sources may arrive from the ceiling. The T_{sub} value was measured in the stall for the source position on the stage, the rear source was longer than the frontal source due to the large volume of the stage house.

7.2 Subjective Preference Judgments

Investigations on historical opera houses with the aim of preserving the cultural heritage were performed based on the proposed guidelines (Pompoli and Prodi 2000). They are useful for rebuilding old theaters that have disappeared and for preservation of existing theaters. A typical opera house is distinguished from a concert hall by its largevolume of stagehouse, orchestra pit, and box seats. Four orthogonal acoustical factors that can fully describe the acoustic sound field in an opera house measurement have been conducted. Similar to previous basic investigation as in the previous chapter, subjective preference test by applying the PCT *in the real space was conducted. The theory of subjective preference allows us to evaluate a sound field in terms of the following four orthogonal acoustical factors (Ando 1985): LL, Δt_1, T_{sub}, and IACC. These factors were identified from the systematic investigation of the sound field by using computer simulation and the listening test (Ando 1998). The subjective preference theory has been validated by tests in real concert halls (Cocchi et al. 1990; Sato et al. 1997). Regarding an opera house, this theory has not been confirmed yet. The present study investigates the relationship between the subjective preference of the sound field and the four orthogonal acoustical factors at each seat.*

7.2.1 Procedure

Romanza "Tormento" by P. Tosti was applied as the source signal. The vocal (soprano) and piano accompaniments were channeled separately. The duration of the signal was 16 s. For music signals containing large fluctuations in tempo and style of performance, the minimum value for effective duration $(\tau_e)_{min}$ of the running ACF of the source signal was applied (Sect. 2.1; Ando et al. 1989; Mouri et al. 2000). The music piece at $(\tau_e)_{min}$ is the most active part, and the listener is sensitive in terms of the changes in the temporal acoustical factors.

Sato et al. (2002) performed subjective preference judgments for the sound field at each seating position in an existing opera house. Vocal and piano signals as source signals applied here were picked up by a microphone in an anechoic chamber, which were reproduced by two loudspeakers. The effective durations of the running ACF were calculated after passing the signal through an A-weighted network. The running integration interval for the ACF, $2T$, was 2.0 s, and the running step was 100 ms. The $(\tau_e)_{min}$ value for the signal mixed was 16 ms as shown in Fig. 7.3. The opera house selected in the experiments was the Teatro Comunale in Ferrara, Italy. The plan of the theater is shown in Fig. 7.2, where the

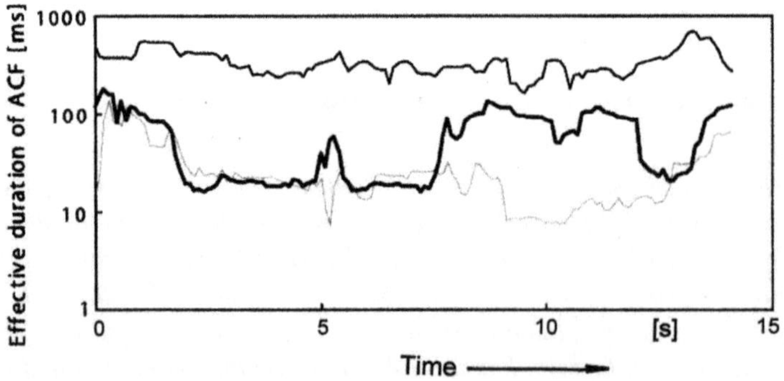

Fig. 7.3 Measured τ_e of the running ACF of the source signal used in the experiment with a 100 ms interval as a function of the time, $2T = 2.0$ s. *Very thin* piano; *thin* vocal; *thick* mixed

positions of sound sources and listeners are also indicated. It has a truncated elliptical plan and consists of 800 seats (two-thirds of them in the five tiers of boxes) with a hall of 5,000 m^2 and a stagehouse of 8,500 m^3. The stage did not contain any scenery set. Curtains were not lowered at the back of the stage. There were no musical instruments nor chairs in the orchestra pit. The pit rail is made of a hard wooden board and is installed between the stall and the orchestra pit. Its height is 2.08 m from the pit floor. The top of the pit rail is in line with the stage.

Two loudspeakers reproducing the vocal signal were located on the stage (one just under the proscenium and the other 2.5 m behind it); and two loudspeakers for reproducing the piano signal were placed in the orchestra pit (one in front of the conductor's box and the other under overhang). The heights of the loudspeakers on the stage and in the orchestra pit were 1.5 and 1.2 m above the floor level, respectively (Fig. 7.1). These heights correspond to the mouth of a singer on the stage and the height of the musical instrument played by a sitting musician in the orchestra pit, respectively.

To obtain reliable subjective responses in an existing sound field, we applied the PCT. This method is simple enough for even nonskilled listeners to judge. To exclude other physical factors such as visual and tactile senses, loudspeakers for source locations on the stage and in the orchestra pit were switched. The PCT using four sound sources in various combinations (Table 7.1) were conducted. The duration of the music signal was 16 s and the silent interval between the stimuli was 2 s. Each pair of sound fields was separated by an interval of 4 s. The tests were performed for all combinations in pairs, i.e., six pairs ($N(N-1)/2$, where $N = 4$) of stimuli for a single session. The pairs were arranged in random order.

7.2.2 Subjects

Forty-seven listeners participated in the experiments. Twenty-one of them were students of the musical department and twenty-seven listeners were students from the Faculty of Engineering. The listeners were divided into ten groups and were seated at specific seats or near them (Fig. 7.2). Five groups sat in the stalls and the other five groups sat in the boxes or gallery. As the source locations of the music changed, listeners were asked to give their acoustical preferences. They were advised to judge every pair (not to leave a blank), facing the center of the stage, and not to copy the answers of other participants. They were also asked to write down their name, age, sex, and musical experience (period and instruments) on their answer sheets. Prior to the experimental sessions, a practice session was conducted by presenting three pairs of stimuli. The experimental session was repeated five times, each time the listeners changed their seats. It took about four minutes for a single session and about thirty minutes in total, including the time for changing seats, and 25 or 26 listeners in total responded at each listener's position.

7.2.3 Results of the Paired-Comparison Tests (PCT)

Results were investigated by tests of consistency of whether the listeners could discriminate the sound fields presented; for example, if a listener prefers sound field A to B, B to C, and C to A. The number of listeners who showed a significant ability to discriminate preferences was 20 in the stalls and 17 in the boxes or gallery. The test of agreement indicated that there was a significant ($p < 0.05$) degree of agreement among the listeners. Scale values of preference were obtained by applying the law of comparative judgment and reconfirmed by goodness of fit (Thurstone 1927; Mosteller 1951). According to the listener's musical experience, the data were grouped into categories and were analyzed, but no significant difference among the group was found. Figure 7.4 shows the scale values of preference for each listening position. The results show that the listeners in the stall preferred the frontal source position on the stage (Conditions 1 and 3). The range of scale values in the boxes was smaller than that of stalls.

7.3 Multiple Dimensional Analyses

The relationship between scale values of subjective preference and the four orthogonal factors and in addition τ_{IACC} obtained by acoustical measurements in an opera house were examined by the multiple factor analysis (Hayashi 1952, 1954a, b; see also, Appendix I, Ando 1998). We may obtain the numeric value to each sub-category of each factor to maximize the correlation coefficient between the scale values of subjective preference and the total scores.

7.3.1 Correlation Matrix of Physical Factors

The scale value of subjective preference obtained at seating positions ① through ⑩ was obtained by the judgments of 47 subjects in total as shown in Fig. 7.4a, b.

Matrix of correlation coefficients between these physical factors are listed in Table 7.2. There is an apparent high coherence between physical factors, for example, LL and Δt_1 of the sound field in the existing opera house. This coherence is a physical phenomenon that depends on the characteristics of the sound field. Thus, each physical factor is ensured as of independent or orthogonal.

The outside variable to be predicted was the scale values of subjective preference obtained by the PCT, and the explanatory variables were: (1) LL, (2) IACC, (3) τ_{IACC}, and (4) Δt_1 for the pit source, (5) Δt_1 for the stage source, (6) T_{sub} of 1 kHz for the pit source, and (7) T_{sub} of 1 kHz for the stage source.

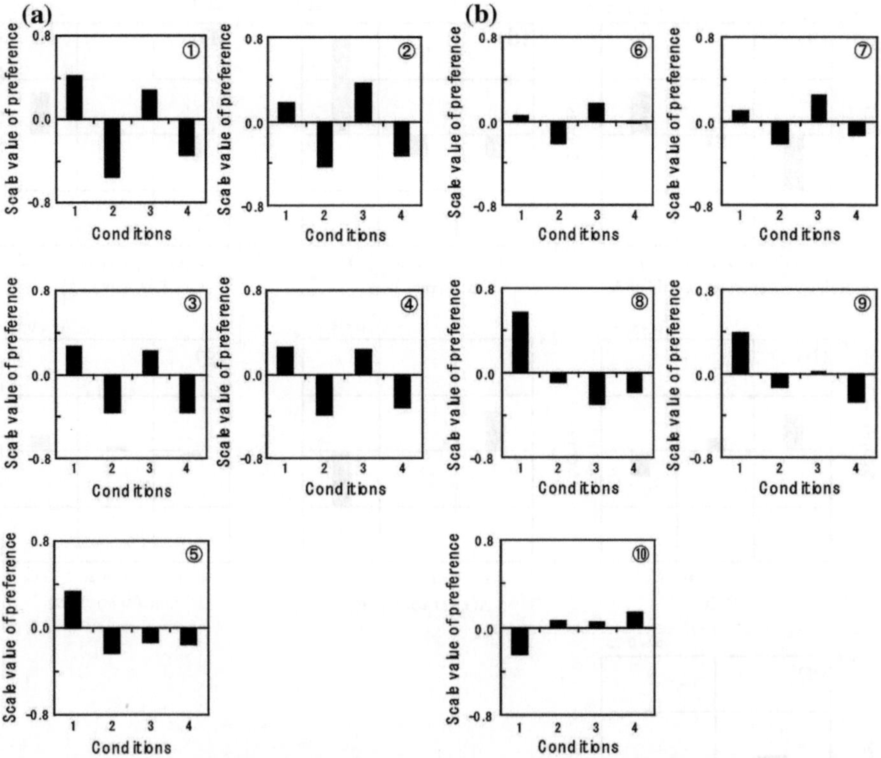

Fig. 7.4 Scale values obtained by paired-comparison tests for four conditions at each listener's location. **a** Stalls ①–⑤. **b** Boxes ⑥–⑨, and gallery ⑩

Table 7.2 Correlation coefficients between physical factors measured in the opera house

Spatial and temporal factors	LL	IACC	τ_{IACC}	Δt_1 (pit)	Δt_1 (stage)	T_{sub} (1 kHz; pit)	T_{sub} (1 kHz; stage)
LL	–	0.12	−0.04	0.74*	0.56*	0.10	−0.17
IACC		–	0.23	0.22	0.25	0.23	0.61*
τ_{IACC}			–	0.02	−0.22	0.13	−0.02
Δt_1 (pit)				–	0.44	0.27	−0.12
Δt_1 (stage)					–	−0.08	−0.06
T_{sub} (1 kHz; pit)						–	0.02
T_{sub} (1 kHz; stage)							–

*$p < 0.01$

7.3.2 Results and Discussion

The scores which give the best correlation between the scale value of preference and the total score obtained from the factor analysis are shown in Fig. 7.5. It is clear

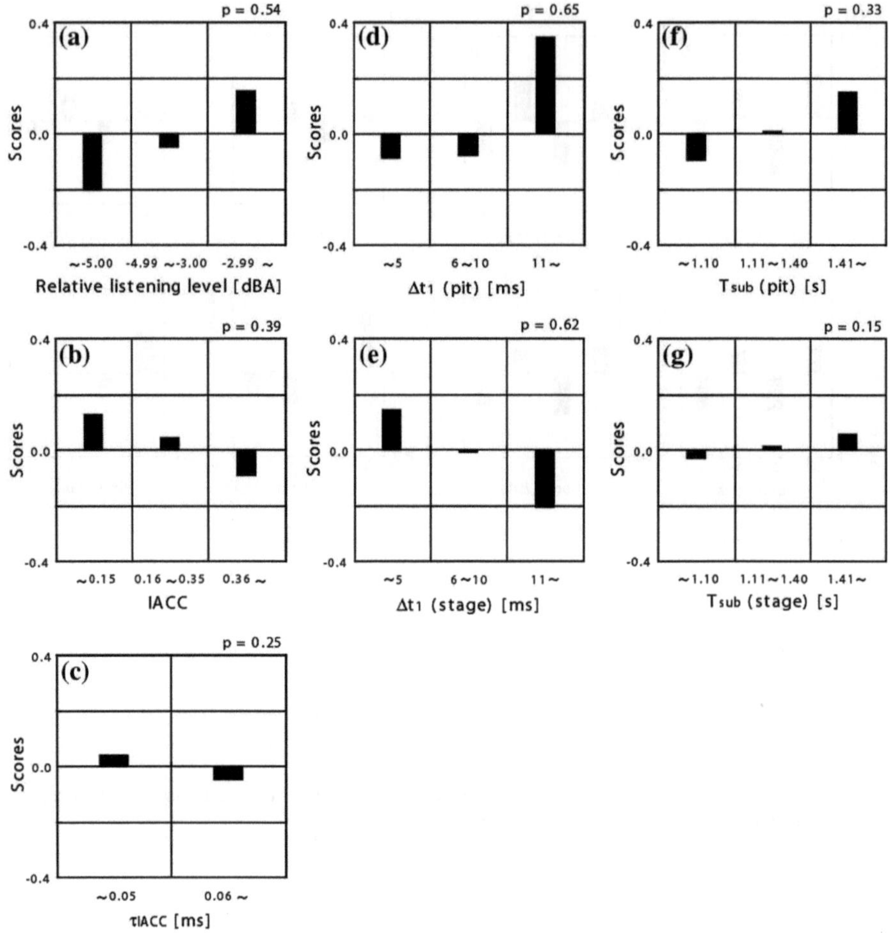

Fig. 7.5 Scores for each category of physical factors obtained by the factor analysis. **a** LL. **b** IACC. **c** τ_{IACC}. **d** Δt_1 for the pit source. **e** Δt_1 for the stage source. **f** T_{sub} of 1 kHz for the pit source. **g** T_{sub} of 1 kHz for the stage source (*p* partial correlation coefficient of each factor)

that the LL scores increased with increasing the LL. The scores decreased with an increase in IACC as described by the theory in the previous chapter. The τ_{IACC} scores slightly decreased with increasing τ_{IACC} due to image shift of sound source as similar to results obtained in an existing concert hall (Sato et al. 1997). The effect of τ_{IACC} on the total scores, however, was minor because all of the loudspeakers were located on the center axis of the hall and the listeners faced the center of the stage ($\tau_{IACC} \sim 0$). The scores for the three factors (LL, IACC, and τ_{IACC}) agreed with those obtained for sound fields in a concert hall also (Chap. 6, Ando 1985).

The contribution to the score as expressed by the partial correlation coefficient of Δt_1 is the largest among the physical factors. The Δt_1 scores for the pit source

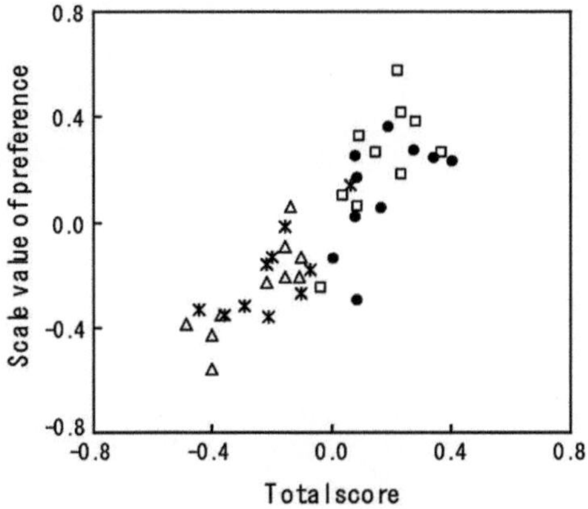

Fig. 7.6 Relationship between scale values obtained by subjective judgments and total scores obtained by factor analysis using scores shown in Fig. 7.5 (*square* condition 1, *triangle* condition 2, *asterisk* condition 3, *bullet* condition 4)

increased with an increase in Δt_1. On the other hand, the Δt_1 scores for the stage source increased with a decrease in Δt_1. Preferred Δt_1 of the source signal with longer effective duration of ACF $(\tau_e)_{min}$ is longer than that with shorter $(\tau_e)_{min}$, as described in Chap. 6. The piano signal reproduced from the loudspeakers in the pit has longer $(\tau_e)_{min}$ ($\approx$160 ms) than the vocal signal ($\approx$8 ms) from the loudspeakers on the stage as shown in Fig. 7.1. The scores of Δt_1 may be related to the $(\tau_e)_{min}$ values of the source signals.

The T_{sub} scores in the pit increased with increasing T_{sub} due to a longer $(\tau_e)_{min}$ ($\approx$160 ms); note that the most preferred calculated $[T_{sub}]_p$ by Eq. (6.8) is about 3. 7 s. The effects of T_{sub} on the scores for the stage vocal source ($[T_{sub}]_p \sim 0.2$ s), however, are apparently to be rather minor in this investigation. The reason for this is the limited range of $T_{sub} \sim$ (1.0–1.5 s) because of the single opera house.

The remarkable results are as follows:

(1) Behavior of scores obtained by factor analysis indicated in Fig. 7.5 in the existing opera house fairly agreed with the results of scale values obtained for simulated sound fields as described in the previous chapter. There are no fundamental contradictions between the theoretical and experimental values at all.
(2) The scale values obtained by preference judgments at each listening position and the total scores added resulting each score (Fig. 7.5) at each listening position agreed closely as shown in Fig. 7.6 ($r = 0.86$, $p < 0.01$).

Chapter 8
Reverberance of the Sound Field

As an application of subjective preference theory with four orthogonal factors described in Chap. 6, this chapter discusses reverberance of the sound field. It shall be shown that the scale value of reverberance obtained by applying speech and music signals may well be expressed commonly by the four orthogonal factors. As similar manner to the scale value of subjective preference investigated for simulated sound fields, we now can calculate those of reverberance at every seating position in an opera house at design stage.

8.1 Reverberance in Relation to Four Orthogonal Factors

8.1.1 Scale Value of Reverberance in Relation to Δt_1 and T_{sub}

First of all, effects of the temporal factors Δt_1 and T_{sub} associated with the left cerebral hemisphere on reverberance were examined (Hase 2001). The values of Δt_1 were changed as 10, 20, and 40 ms, and the values of T_{sub} as 0.25, 0.85, and 1.25 s. In addition, after examination of effects of different source signals, we applied speech and music signals (5 s pieces) with different minimum values of effective duration of the running ACF as listed in Table 8.1. The values of IACC and SPL were fixed at nearly preferred conditions 0.40 and 82 dBA at a peak level, respectively.

The paired comparison tests (PCT) with five subjects (36 pairs) were conducted. The session was repeated 10 times for each subject. Results of the PCT with all of five subjects are shown in Fig. 8.1 as a function of Δt_1 and a parameter of T_{sub}. The result of analysis of variance for the scale value of reverberance is indicated in Table 8.2. Effects of Δt_1 and T_{sub} on reverberance were independent and significant on the scale value and no residual was significant, so that

$$S(\text{rev.})_{\text{L}} \approx a_2 x_2 + a_3 x_3 \tag{8.1}$$

Y. Ando, *Opera House Acoustics Based on Subjective Preference Theory*, Mathematics for Industry 12, DOI 10.1007/978-4-431-55423-3_8

Table 8.1 Source signals used in experiment

Source signal	Title	$(\tau_e)_{min}$ (ms)	$[T_{sub}]_p$ calculated (s)
Music (9 s)	Marriage of Figaro overture (by Mozart)	59	1.4
Speech (9 s)	Female speech (Japanese)	11	0.3

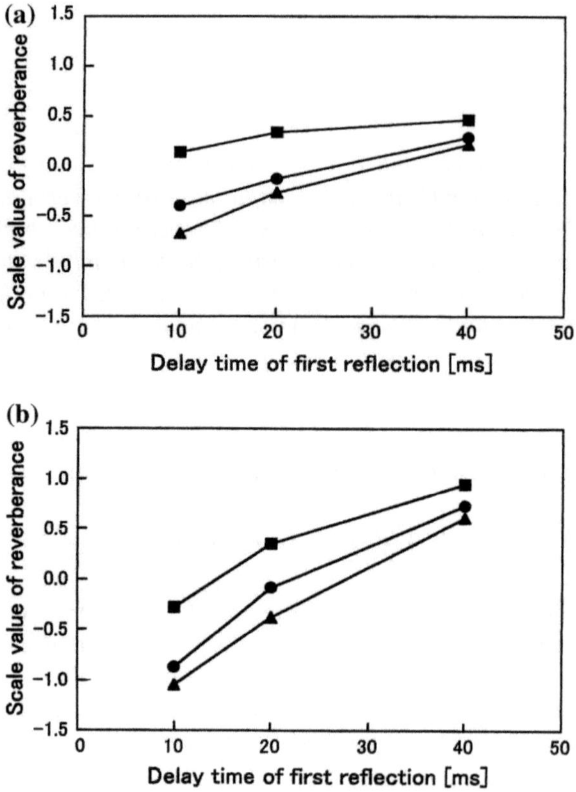

Fig. 8.1 Averaged scale value of reverberance as a function of Δt_1 and as a parameter of T_{sub} (5 subjects). **a** Music. **b** Speech. *Filled triangle* T_{sub} = 0.25 s; *Filled circle* T_{sub} = 0.85 s; *Filled square* T_{sub} = 1.25 s

Table 8.2 Results of the analysis of variance for the scale values of reverberance

(a) Music					
Factors	Sum of square	DF	Mean square	*F*-ratio	*p*-value
Δt_1	6.0	2	3.0	40.0	<0.01
T_{sub}	3.9	2	1.9	25.7	<0.01
Residual	2.7	36	0.1		
(b) Speech					
Factors	Sum of square	DF	Mean square	*F*-ratio	*p*-value
Δt_1	15.3	2	7.7	161.8	<0.01
T_{sub}	2.4	2	1.2	25.8	<0.01
Residual	1.8	45	0.1		

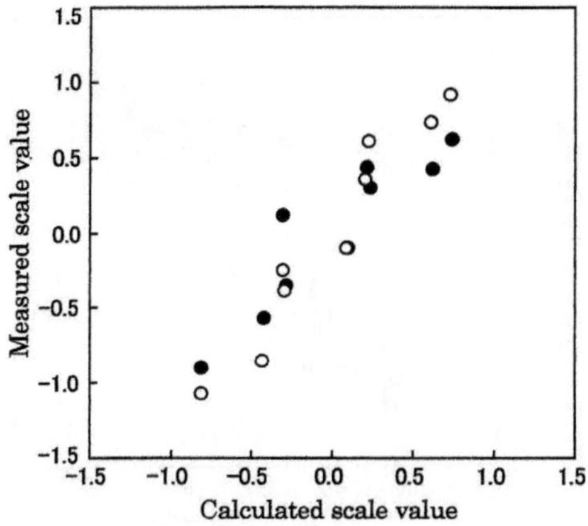

Fig. 8.2 Relationship between the measured scale values of reverberance and the scale values calculated by Eq. (8.1) with $a_2 = 1.75$ and $a_3 = 0.71$. The correlation coefficient, $r = 0.93$ ($p < 0.01$). *Filled circle* music; *open circle* speech

Table 8.3 Coefficients a_2 and a_3 in Eq. (8.1) for calculating reverberance of each individual, and the correlation coefficient between the measured and calculated scale values

Individual	a_2	a_3	Correlation coefficient
A	1.54	0.96	0.81
B	1.53	1.07	0.93
C	1.87	0.55	0.86
D	2.03	0.58	0.95
E	2.08	0.44	0.93
Average	1.75	0.71	0.93

where x_2 and x_3 are the normalized temporal factors defined by Eqs. (6.13) and (6.14), respectively. Suffix L of S(rev.) signifies specialization associated with the left hemisphere. Coefficients a_2 and a_3 averaged for five subjects were 1.75 and 0. 71, respectively. Figure 8.2 shows the resulting relationship between measured scale values and calculated scale values by Eq. (8.1), $r \approx 0.93$ ($p < 0.01$). For each individual, coefficients a_2 and a_3 obtained are listed in Table 8.3, and relationships between individual measured scale values and calculated values by Eq. (8.1) with individual coefficients are shown in Fig. 8.3, $r \approx 0.93$ ($p < 0.01$).

Calculated scale values by Eq. (8.1) fairly agree with measured ones for both speech and music signals. It is remarkable that contributions of the factors of Δt_1 with the speech signal were much greater than those of T_{sub} for every individual in the range of this experiment, but this tendency was not clear with music signal.

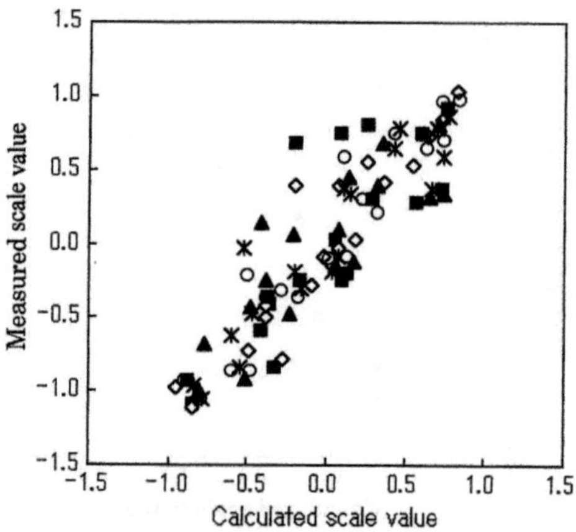

Fig. 8.3 Relationship between the measured individual scale values of reverberance and the scale values calculated by Eq. (8.1) with individual coefficients a_2 and a_3 listed in Table 8.3. The correlation coefficient, $r = 0.90$ ($p < 0.01$). *Different symbols* are values for different subjects

8.1.2 Scale Value of Reverberance in Relation to SPL and IACC

Second, in order to examine effects of the spatial factors associated with the right cerebral hemisphere, SPL and IACC were varied (Hase 2001). The value of SPL was controlled as 76, 82, and 88 dBA, and the value of IACC was controlled by changing horizontal angle of two early reflections $\xi = \pm 4°$, $\pm 22°$, and $\pm 54°$, so that IACC = 0.60, 0.45, and 0.30 (music signal); IACC = 0.75, 0.60, and 0.40 (speech signal), respectively. In order to examine the effects of different source signals here also, we applied the same two signals of speech and music signals as previous section (Table 8.1). The values of Δt_1 and T_{sub}, respectively, were fixed at conditions in the range of most preferred value calculated by Eqs. (6.7) and (6.8), so that 41 ms and 1.3 s (music signal); 8 ms and 0.3 s (speech signal). The reverberation time was controlled by the loudspeaker fixed at $\xi = 0°$ that was the same location for the direct sound.

The PCT with nine subjects (36 pairs) were conducted. The session was repeated 10 times for each subject. Results of the PCT with all of nine subjects are shown in Fig. 8.4 as a function of the IACC and a parameter of SPL. The result of analysis of variance for the scale value of reverberance is indicated in Table 8.4. Effects of LL and IACC are independent and significant on the scale value and not residual, so that

$$S(\text{rev.})_R \approx a_1 x_1 + a_4 x_4 \qquad (8.2)$$

where x_1 = SPL − 82 (dBA) in this experimental condition as similar to Eq. (6.12), and x_4 = IACC, same as Eq. (6.15). Suffix R of S(rev.) signifies specialization associated with the right hemisphere. Coefficients averaged a_1 and a_4 obtained by

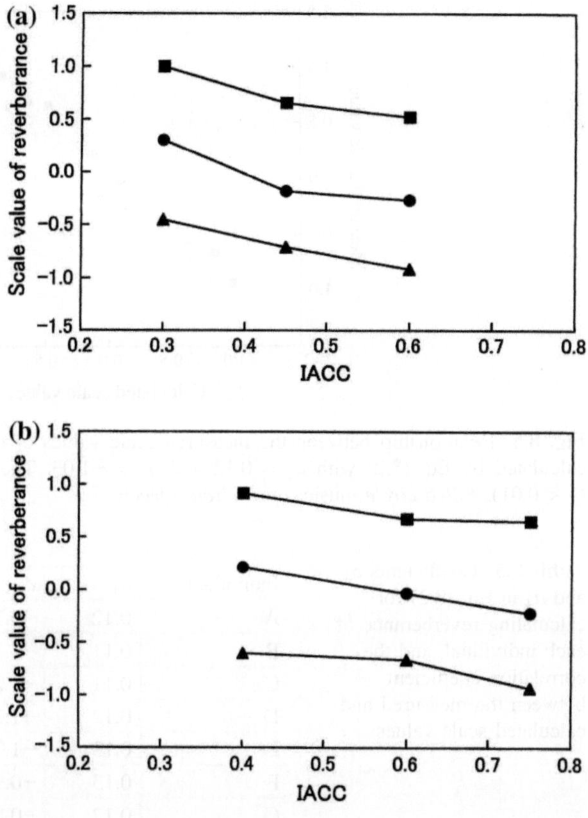

Fig. 8.4 Averaged scale values of reverberance as a function of IACC and as a parameter of LL (9 subjects). **a** Music. **b** Speech. *Filled triangle* LL = 76 dBA; *Filled circle* LL = 82 dBA; *Filled square* LL = 88 dBA

Table 8.4 Results of the analysis of variance for the scale values of reverberance

(a) Music					
Factors	Sum of square	DF	Mean square	*F*-ratio	*p*-value
LL	29.4	2	14.7	886.9	<0.01
IACC	1.6	2	0.8	48.1	<0.01
Residual	1.2	72	0.02		
(b) Speech					
Factors	Sum of square	DF	Mean square	*F*-ratio	*p*-value
LL	26.7	2	13.3	702.3	<0.01
IACC	3.9	2	1.9	101.1	<0.01
Residual	1.4	72	0.02		

nine subjects were 0.12 and −1.03, respectively. Figure 8.5 shows the relationship between measured scale values and calculated scale values by Eq. (8.2), $r \approx 0.97$ ($p < 0.01$). For each individual, coefficients a_1 and a_4 obtained are listed in Table 8.5.

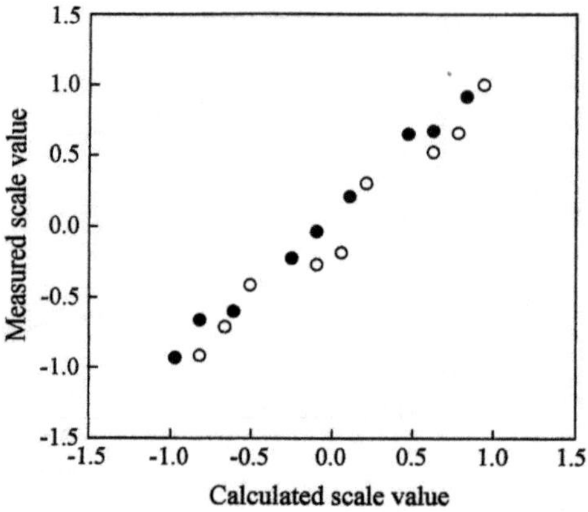

Fig. 8.5 Relationship between the measured scale values of reverberance and the scale values calculated by Eq. (8.2) with a_1 = 0.12 and a_4 = −1.03. The correlation coefficient, r = 0.98 ($p < 0.01$). *Filled circle* music; *open circle* speech

Table 8.5 Coefficients a_1 and a_4 in Eq. (8.2) for calculating reverberance of each individual, and the correlation coefficient between the measured and calculated scale values

Individual	a_1	a_4	Correlation coefficient
A	0.12	−0.70	0.98
B	0.11	−1.38	0.93
C	0.11	−1.27	0.96
D	0.12	−1.35	0.96
E	0.13	−1.20	0.98
F	0.13	−0.94	0.98
G	0.12	−0.86	0.98
H	0.13	−0.34	0.98
I	0.12	−1.27	0.98
Average	0.12	−1.03	0.98

Resulting relationships between individual measured scale values and calculated values by Eq. (8.2) with individual coefficients are shown in Fig. 8.6, $r \approx 0.98$ ($p < 0.01$).

Calculated scale values by Eq. (8.2) fairly agree with measured ones for both speech and music signals. The scale value of reverberance increases with increasing LL and decreasing IACC. It is remarkable that no significant differences may be found in the difference of source signals, due to the spatial factors varied.

Since activities in the left and right cerebral hemispheres are independent, we assume that Eqs. (8.1) and (8.2) may be combined together obtaining overall response of reverberance, such that

$$S(\text{rev.}) = S(\text{rev.})_{\text{L}} + S(\text{rev.})_{\text{R}} \approx a_1x_1 + a_2x_2 + a_3x_3 + a_4x_4 \quad (8.3)$$

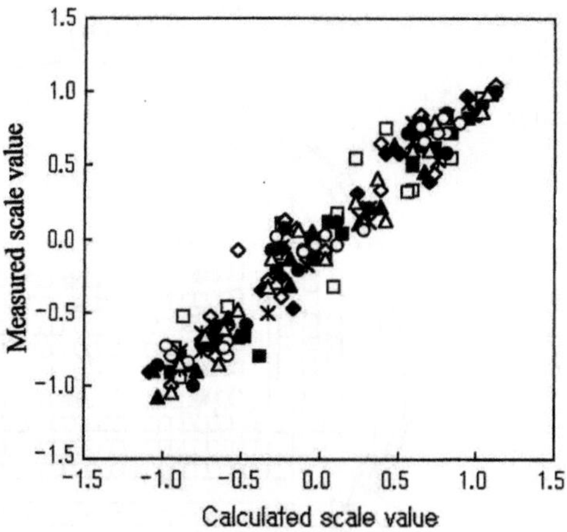

Fig. 8.6 Relationship between the measured individual scale value of reverberance and the scale value calculated by Eq. (8.2) with individual coefficients a_1 and a_4 listed in Table 8.5. The correlation coefficient, $r = 0.97$ ($p < 0.01$). *Different symbols* are values for different subjects

where $a_1 \approx 0.12$, $a_2 \approx 1.75$, $a_3 \approx 0.71$, and $a_4 \approx -1.03$. Thus, the scale value of reverberance at each seating position might be calculated by Eq. (8.3) as similar manner to that of subjective preference by Eq. (6.10). We shall examine this in an existing room as discussed in the following section.

8.2 Examination on Reverberance in an Existing Hall

In order to examine Eq. (8.3) *resulted above, we controlled SPL (LL) as one of spatial factors and T_{sub} as one of temporal factors in an actual existing hall.*

The sound field at the seating positions in the Orbis Hall (with 400 seats), Kobe (Fig. 8.7) was applied (Hase et al. 2000). Music and speech were applied also as a source signal as listed in Table 8.1. The minimum value of the effective duration $(\tau_e)_{min}$ of the running ACF was obtained with the integration interval $2T = 2.0$ s. The PCT was conducted as SPL (LL) and T_{sub} were changed. The value of T_{sub} was adjusted by a hybrid system, consisting of an electroacoustic system and a small reverberation chamber, which reproduced fine structured reflections in the decay, as shown in Fig. 8.8. Sound signals were reproduced from an omnidirectional dodecahedron loudspeaker located on the stage (1.35 m above the stage floor, and 2.20 m from the front edge of the stage) and picked up by a microphone on the stage and fed into the hybrid system (0.50 m from the loudspeaker and 1.10 m above the stage floor). The signals radiated from the loudspeakers distributed near the ceiling were delivered through the hybrid reverberator and were superposed on the sound field in the hall. Averaged values of T_{sub} with and without the reverberator are listed in Table 8.6. Because there was little difference in the measured values of T_{sub} among seats, the measured values at the six seat positions were

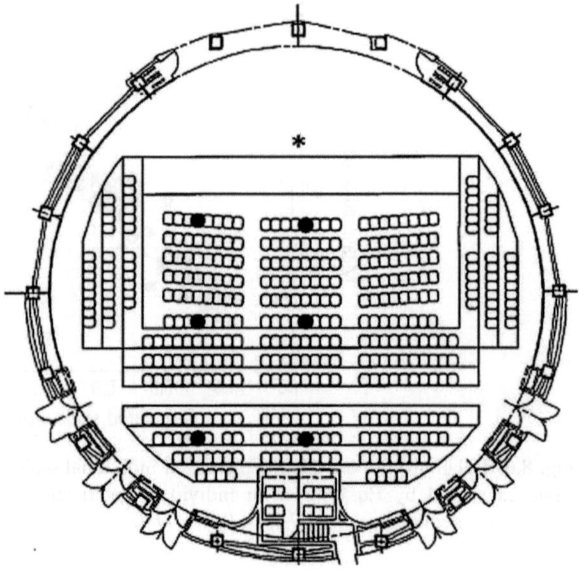

Fig. 8.7 Plan of the existing hall (ORBIS HALL, Kobe). *Star* sound source; *filled circle* listener's location (6 seating positions)

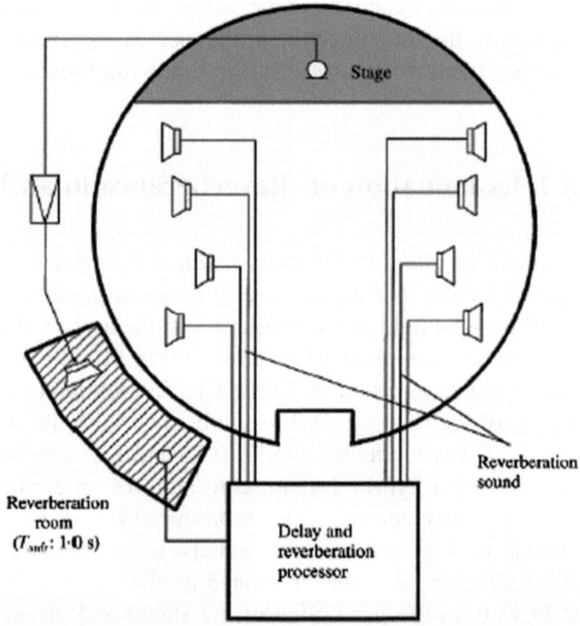

Fig. 8.8 Block diagram of a hybrid reverberation system

averaged. The frequency characteristics of measured T_{sub} without reverberation system was almost flat around 1.0 s. When the reverberation was superposed, on the other hand, the averaged value of T_{sub} was increased to about 1.4 s.

Four conditions of the experiment are listed in Table 8.7. Twenty-one subjects were divided into six groups and seated at the specified positions as shown in

Table 8.6 Range T_{sub} controlled at six seats

	T_{sub} (s)						
Reverberator	125 Hz	250 Hz	500 Hz	1 kHz	2 kHz	4 kHz	Averaged
Off	1.0	0.9	0.9	1.0	1.2	1.1	1.0
On	2.2	1.2	0.9	1.3	1.7	1.4	1.4

Note Off: without the reverberation system; on: with the reverberation system

Table 8.7 Four conditions of the experiment

Factors	1	2	3	4
SPL (LL)	80 dB	80 dB	70 dB	70 dB
T_{sub}* (s)	1.0	1.5	1.0	1.4

Fig. 8.7. To avoid effects of other senses including visual and tactile senses on judgments, the subjects were asked to remain in their seats (Sato et al. 1997), and judge which of two sound fields they perceived to have more reverberance. The test consisted of six pairs ($N(N-1)/2$, $N = 4$) of stimuli, in total. The signal duration of each stimulus was 9 s, and the silent interval between the stimuli was 1 s. Each pair of sound fields was separated by an interval of 4 s and the pairs were arranged in random order. The session was repeated three times with subjects changing seats between sessions.

The scale value of reverberance was obtained by applying the method, a modification of the Thurstone method (1927). Because there was no significant difference in the scale value of reverberance among the seats in the existing hall, scale values of the six groups of seats were averaged. As shown in Fig. 8.9, the scale

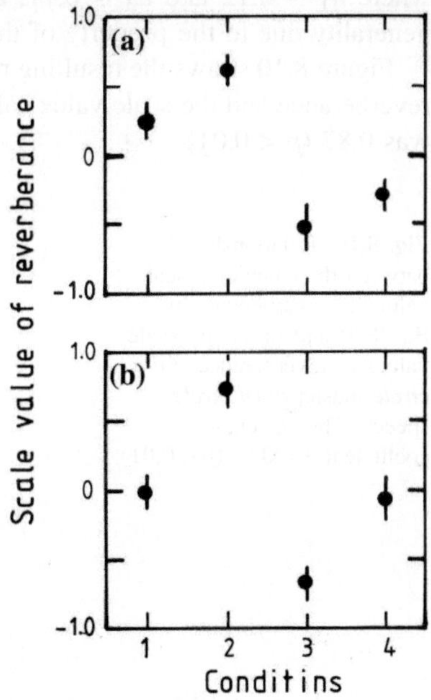

Fig. 8.9 Average scale values of reverberance (6 seating positions) with 95 % reliability. **a** Music. **b** Speech. *Note* On: with the reverberation system; off: without the reverberation system

Table 8.8 Results of the analysis of variance for the scale values of reverberance

(a) Music					
Factors	Sum of square	DF	Mean square	F-ratio	p-value
LL	4.43	1	4.4	312.0	<0.01
T_{sub}	0.47	1	0.5	32.9	<0.01
Seat	0.00	5	0.0	0.0	
LL and T_{sub}	0.03	1	1.8	1.8	
Residual	0.07	5	0.0		
(b) Speech					
Factors	Sum of square	DF	Mean square	F-ratio	p-value
LL	3.29	1	3.3	72.6	<0.01
T_{sub}	2.69	1	2.7	59.5	<0.01
Seat	0.00	5	0.0	0.0	
LL and T_{sub}	0.02	1	0.0	0.5	
Residual	0.23	5	0.1		

value of reverberance for the sets of both sound signals increased as SPL or T_{sub} increased. Results of the analysis of variance for the scale values of reverberance indicate that the factors SPL and T_{sub} are significant ($p < 0.01$), and the interaction between SPL and T_{sub} are not significant (Table 8.8). Thus, we can superpose the scale value due to two factors only in Eq. (8.3), so that

$$S(\text{rev.}) = S(\text{rev.})_L + S(\text{rev.})_R \approx a_1 x_1 + a_3 x_3 + (\text{const.}) \tag{8.4}$$

where $a_1 \approx 0.12$ and $a_3 \approx 0.71$, and (const.) are eliminated without losing any generality due to the property of the scale value.

Figure 8.10 shows the resulting relationship between the measured scale value of reverberance and the scale value calculated by Eq. (8.4). The correlation coefficient was 0.87 ($p < 0.01$).

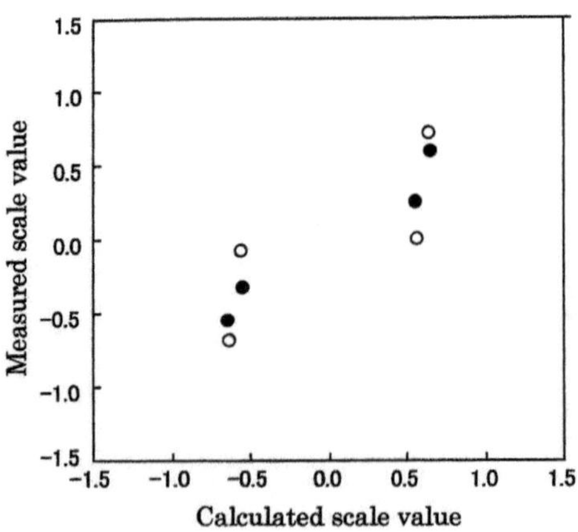

Fig. 8.10 Relationship between the calculated scale values of reverberance by Eq. (8.4) and measured scale values of reverberance. *Filled circle* music; *open circle* speech. The correlation coefficient, $r = 0.87$ ($p < 0.01$)

So far Eq. (8.3) may be roughly reconfirmed by the bridge of the two hemispheric independent activities as discussed in Chaps. 6–8. Effects of the SPL (LL) on reverberance can be related to the sensation level of reverberation decay, so that the higher SPL could result in greater values of reverberance. Also, source signals with the shorter $(\tau_e)_{min}$ value may result more reverberance.

Theory of subjective preference (Sect. 6.6), therefore, may be applied "fundamentally" for reverberance as an overall response of the sound field, also.

Chapter 9
Improvements in Subjective Preferences for Listeners and Performers

9.1 Effects of Stage Building of Ancient Theaters

Acoustical measurements were conducted in two ancient theaters—one Greek and one Roman—the origin of modern opera houses. The Delphi theater (Greece) does not have a stage building, while the Taormina theater (Italy) has a partially remaining stage building behind the orchestra. The effect of the stage building on the sound field was determined in terms of the temporal and spatial factors. Typically, the complicated stage building has affects on improving the values of T_{sub} *and IACC.*

9.1.1 Binaural Impulse Responses

Acoustical measurements were conducted by Sato et al. (2002) in the Delphi theater (5,000 seats) and the Taormina theater with 5,400 seats. The Delphi theater does not have a stage building as shown in Fig. 9.1a, while the Taormina theater has a complicated stage building behind the orchestra as can be seen in Fig. 9.1b. Many of the seats of the Delphi theater were carved out of rock. In the Taormina theater, the frontal section of seating (M01, M04, and M07) consists of seating planks on a temporary steel frame that has a high degree of acoustic transparency. The middle seating area (M02, M05, and M08) consists of cut stone, and the rear (M03, M06, and M09) consists of wood benches.

The test signal was a log sine sweep with a duration 20 s (sampling frequency: 48 kHz). Frequency range was from 80 to 18 kHz. The log sine sweep was radiated from an omnidirectional loudspeaker located at the center of the orchestra. A human head with two small condenser microphones at each ear entrance was used as a receiver. During the measurements, the stage was completely empty. The measurements were conducted in an unoccupied condition. All the measurement devices

Y. Ando, *Opera House Acoustics Based on Subjective Preference Theory*,
Mathematics for Industry 12, DOI 10.1007/978-4-431-55423-3_9

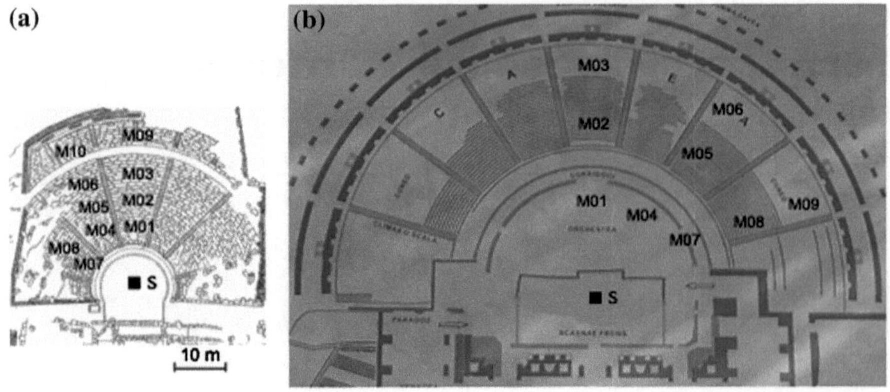

Fig. 9.1 Delphi theater (5,000 seats) and Taormina theater with 5,400 seats

were controlled by a laptop personal computer (PC). Orthogonal physical factors in accordance with the subjective preference theory of a sound field were analyzed.

Figure 9.2 shows binaural impulse responses as typical examples of the measurements. In the Delphi theater, only the single reflection (Δt_1) $\sim$ 2 ms from the orchestra floor was observed at the receiver positions except for the side seat. Strong reflections of about 40 ms were observed in the side seats (M07 and M08).

In the Taormina theater, the delay time of first reflections (Δt_1) $\sim$ 30–40 ms were observed at the seat in the middle row (M02, M05, and M08).

9.1.2 Reverberation

Examples of measured reverberation curve at 500 Hz are shown in Fig. 9.3. Both are open theaters without any diffusing element in the sound field, therefore, the decay has a nonexponential behavior below 250 Hz. It can be seen that the decay of the Delphi theater is steeper than that of the Taormina theater. The sound field decay of the Delphi theater begins immediately after the reflection from the floor. The acoustical decay is sustained by scattering from the stone seats. Measured subsequent reverberation time T_{sub} is shown in Fig. 9.4. Linear regression for initial 20-dB attenuation is calculated by a logarithmic transformation of the integrated decay curve.

Reflections from the orchestra floor and the stage tower play an important role for scattered and reverberant sound fields. For the Delphi theater, values of T_{sub} were 0.5–0.6 s at mid frequencies (500 Hz and 1 kHz, averaged), and values of T_{sub} of Taormina were 0.9–1.0 s.

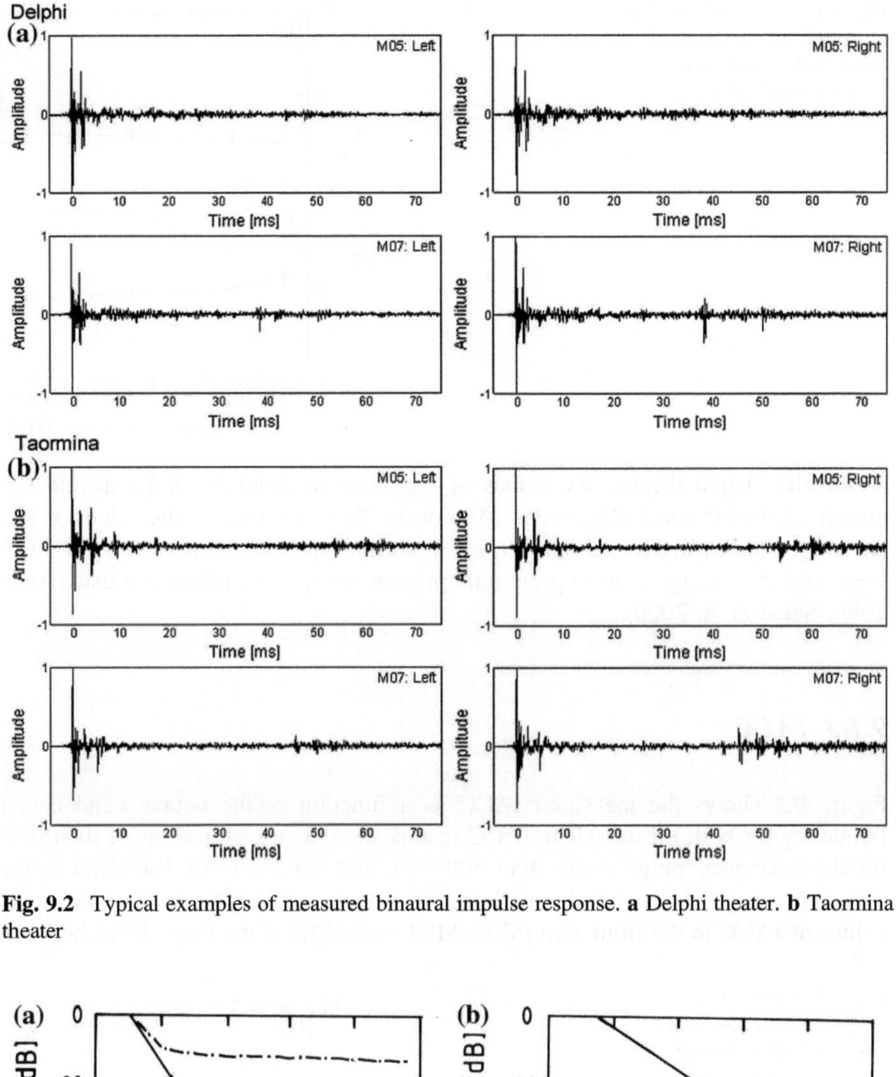

Fig. 9.2 Typical examples of measured binaural impulse response. **a** Delphi theater. **b** Taormina theater

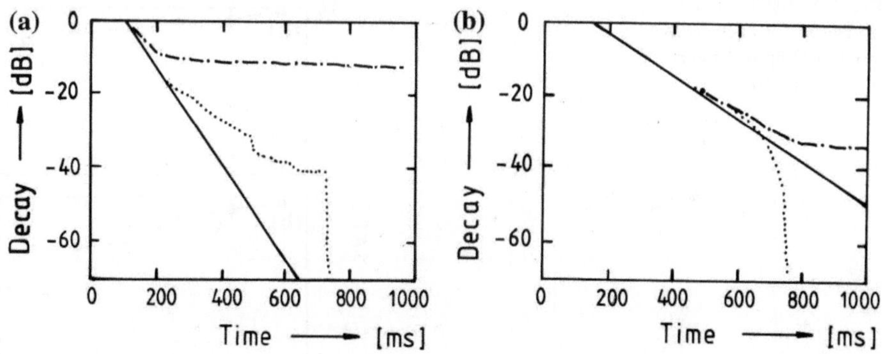

Fig. 9.3 Examples of measured reverberation curve for the octave band centered on 500 Hz. **a** Delphi theater. **b** Taormina theater

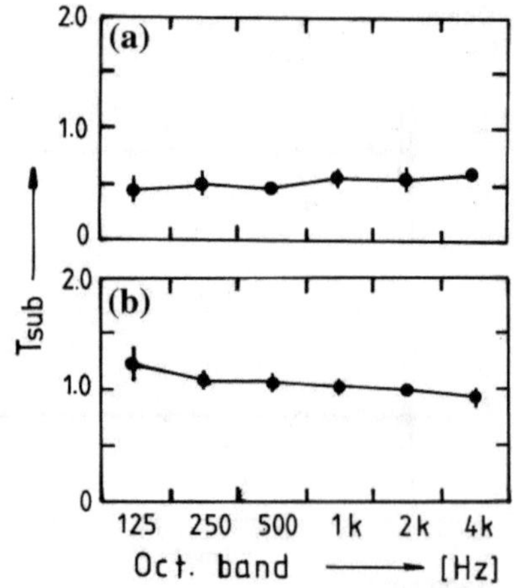

Fig. 9.4 Reverberation time measured for the octave bands. **a** Delphi theater. **b** Taormina theater

For the Delphi theater, the values of T_{sub} were at about 0.5–0.6 s at mid frequencies (500 Hz and 1 kHz, averaged). For the Taormina theater, the values of T_{sub} are longer increasing to around 0.9–1.0 s, so that the reverberation time is much improved at the range of most preferred optimum for speech and vocal music (Ando 1998; Sakai et al. 2000).

9.1.3 IACC

Figure 9.5 shows the measured IACC as a function of the octave band center frequency for both theaters. The IACC results show a large value (more than 0.85 for the frequency range lower than 500 Hz), and the IACC of Taormina in the frequency range above 1 kHz are lower than those of Delphi. In Taormina, the values of IACC in the front area (M01, M04, and M07) at the frequencies between

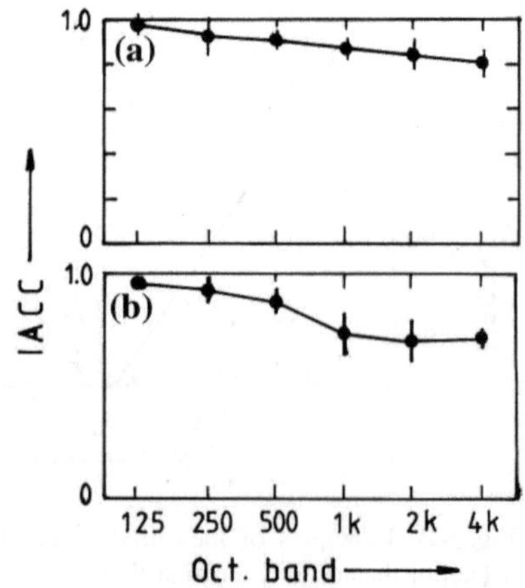

Fig. 9.5 IACC values measured for each octave band. **a** Delphi theater. **b** Taormina theater

250 Hz and 2 kHz were larger than those in the middle and rear areas. Not only the stage building behind the source, but also the masonry walls of the stage sides provide the lateral reflection to decrease the IACC. The stage reflections of Taormina, therefore, made to decrease the IACC.

The values of interaural time delay at IACC were less than 0.10 ms in frequencies greater than 250 Hz at the receiver positions except for the M07 of Taormina theater, thus resulting in a horizontal balance of the sound field.

As concluding remarks, the stage building is improved in both temporal and spatial factors of the sound field.

9.2 Balance of a Vocal Source on the Stage and Music in the Pit of Opera Houses

In order to realize the excellent performance, we shall discuss acoustical balance between a singer on the stage and the orchestra music in the pit inside a historical opera house by applying related orthogonal factors of the sound field.

9.2.1 Balance of Listening Level

Utilizing a set of acoustic data from the Teatro Comunale in Ferrara, the sound field has been controlled by factors related to the balance between the stage and the pit (Prodi and Velecka 2005). Applying an anechoic music piece for soprano with piano accompaniment, listening tests have been conducted in an anechoic chamber equipped with an acoustic system.

Results of the scaling exercise show that the balance of the sound from the stage and the pit has been perceived, when the listening level difference (LLD) between the two is comprised such that

$$-2.0 < \text{LLD} < +2.3 \quad (\text{dBA}) \tag{9.1}$$

9.2.2 Balance of EDT, Δt_1, and IACC

Now we shall discuss effects on reverberance of the early decay time, EDT (deeply related to T_{sub}), the initial time delay gap of the first reflection after arriving the direct sound, Δt_1, and the magnitude of interaural cross-correlation (IACC).

In this study, virtual sound fields developed for a total of seven receiving positions inside a group of four theaters were reproduced by use of the stereo-dipole technique (Kirkeby et al. 1998).

Table 9.1 Three orthogonal factors other than SPL (LL) measured

Opera house	Source position	T_{sub} (s)	EDT (s)	Δt_1 (ms)	IACC
Delphi	Stage	0.28	0.25	28	0.74
	Orchestra	0.35	0.34	14	0.87
Ferrara (stall)	Stage	1.27	0.93	24	0.51
	Pit	1.27	1.15	14	0.10
Ferrara (box)	Stage	1.32	0.94	24	0.14
	Pit	1.10	0.85	29	0.11
Modena (stall)	Stage	1.42	1.35	59	0.33
	Pit	1.28	1.26	40	0.14
Modena (gallery)	Stage	1.46	1.36	11	0.10
	Pit	1.23	1.42	6	0.16
Fernese (orchestra)	Stage	2.86	2.87	11	0.48
	Orchestra	2.73	2.30	13	0.50
Fernese (auditorium)	Stage	2.72	2.80	23	0.10
	Orchestra	2.71	2.74	14	0.12

Binaural impulse responses were measured in open-air and close theaters, which were selected to obtain wide ranges of orthogonal factors other than the listening level. These are listed in Table 9.1. Locations of the sound sources and receivers are shown in Fig. 9.6. The area in front of the stage is known as "orchestra" in ancient theaters.

An exponential sine sweep with variable duration 20–30 s was supplied to the loudspeaker and sound signals were recorded by PC through binaural probes.

An anechoic signal "Tormento" by P. Tosti was applied, the soprano vocal and piano keyboard of 16 s were channeled separately. These signals contained frequency range mainly between 250 and 4,000 Hz.

The sound signals presented in the test were obtained by the following three steps:

(1) *Binaural signals* The binaural impulse responses from the source on the stage or the source in the pit/orchestra were convolved with the anechoic signals. These two signals were mixed at the same amplitudes, i.e., the preferred condition (Nrodi and Velecka 2005).
(2) *Reproduction system* Impulse responses between two loudspeakers and two ears at a distance of 2 m were measured in the test room. The same hardware systems were used in the real theaters. By use of the impulse responses, the respective inverse filters for cross-talk canceling were developed (Figs. 3.1 and 3.2).
(3) *Reproduction of sound fields in real theaters for listening tests* The mixed signals in step 1 were convolved with filters obtained by step 2. Finally, the obtained signals were played back through two loudspeakers to reproduce sound fields.

The accuracies of system in the reproduction of the sound fields examined by means of average absolute differences between the factors in the real and the virtual sound fields were: 0.06 for EDT, 0.06 for T_{sub}, 0.04 for IACC in the six octave

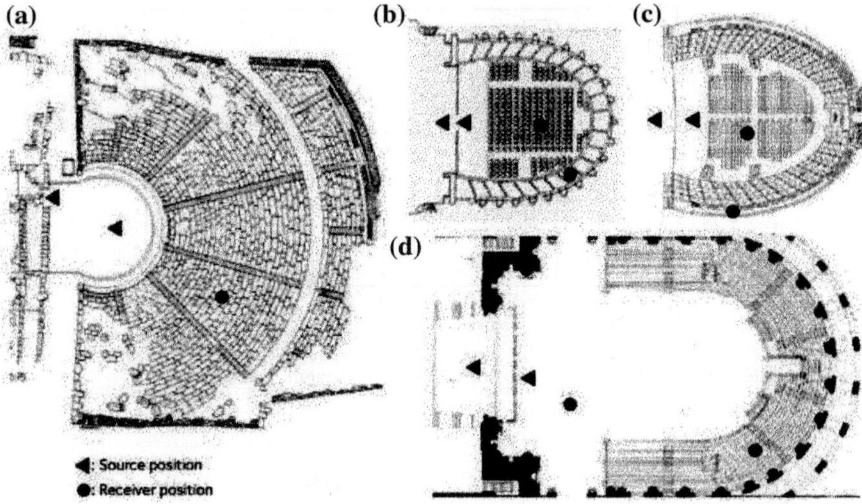

Fig. 9.6 Sound source and receiving positions of binaural impulse response measurements

bands from 125 to 4,000 Hz; these are within just noticeable difference (Prodi and Velcka 2003).

PCTs were conducted for all the combinations, i.e., 21 pairs, ($N(N-1)/2$, where $N = 7$ of stimuli for a single session. The SPL were fixed at 80 dBA, the preferred condition (Ando 1998). For single session of PCTs it was 13.3 min, divided into two parts avoiding fatigue effects.

9.3 Results

As is usual, tests of consistency (Kendall and Smith 1940) were performed and showed that 15 listeners out of 19 showed an ability to discriminate balance, so that the remaining 4 of them were dropped. Also, the test of degree agreement indicated that the degree of agreement in 15 listeners was satisfactory. Then the scale value of balance was obtained (Thurston 1927, Case V).

Figure 9.7 shows the scale values of balance of four theaters with locations of sound source. Next, multiple regression analysis was carried out to obtain effects of the orthogonal factors (Hayashi 1952, 1954a, b, see also Appendix I in Ando 1998). The relationship between the scale values of balance and each factor is shown in Fig. 9.8a, b. Then, the scale values of balance were formed by means of interpolation from a nonlinear equation with the procedure similar to obtaining Eq. (6.11) in Sect. 6.4, so that

$$S = S_1 + S_2 + S_3 \tag{9.2}$$

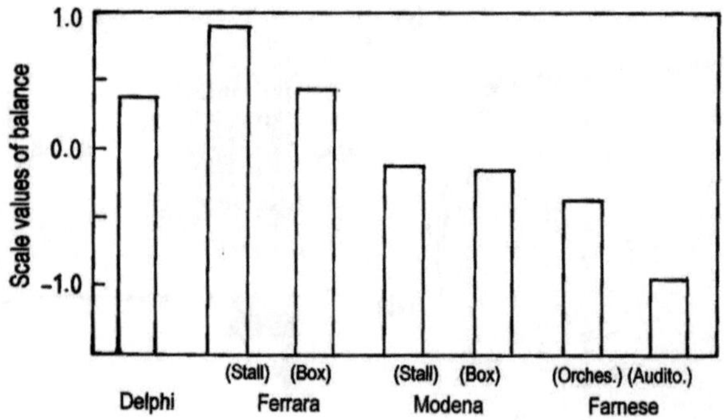

Fig. 9.7 Scale values of balance for the sound fields of theaters investigated

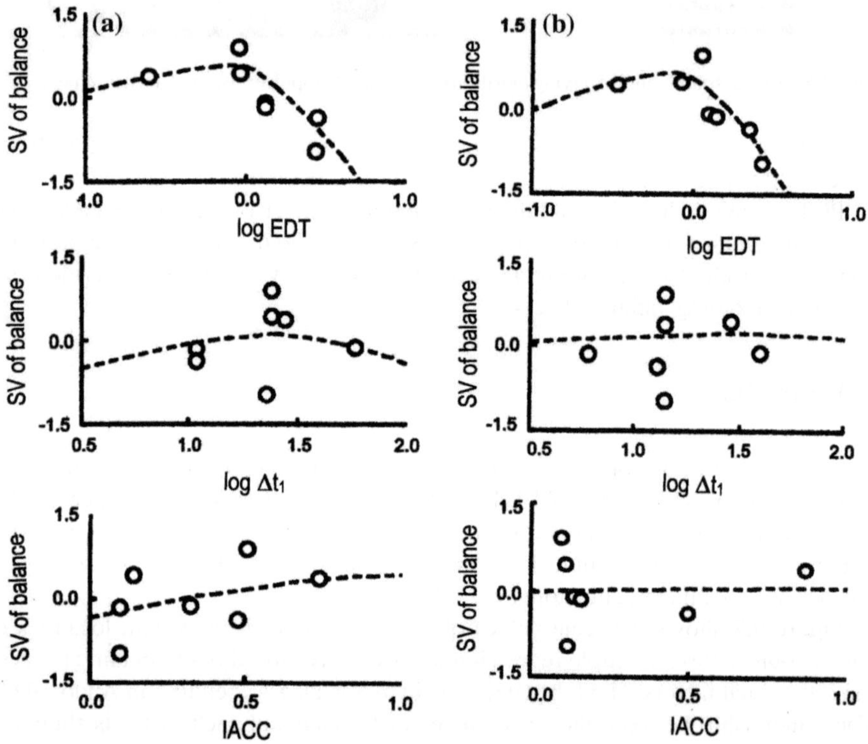

Fig. 9.8 Relationship between the scale value and each of three orthogonal factors *dashed lines* obtained due to Eq. (9.2). **a** Stage sound source. **b** Pit/orchestra sound source

where

$$S_1 = a_1|x_1|^{3/2},\ S_2 = a_2|x_2|^{3/2},\ \text{and}\ S_3 = a_3|x_3|^{3/2}.$$

$$x_1 = \log \mathrm{EDT}/[\mathrm{EDT}]_p \quad (9.3)$$

$$x_2 = \log \Delta t_1/[\Delta t_1]_p \quad (9.4)$$

$$x_3 = \mathrm{IACC} \quad (9.5)$$

The scale values of balance were obtained for orthogonal factors of sound field similar to the subjective preference theory.

The resulting coefficients are listed in Table 9.2a, b. Relationship between the scale values obtained by subjective judgments and the scale values calculated by Eq. (9.2) for both source positions are shown in Fig. 9.9.

Table 9.2 Coefficients and the optimum values of each of three factors in Eq. (9.1) resulted by the experiments

Factor	a_n $(X < 0)$	a_n $(X > 0)$	Optimum value $[X]_p$
(a) Stage source			
EDT (n = 1)	−0.50	−2.89	0.85 (s)
Δt_1 (n = 2)	−0.68	−1.24	27 (ms)
IACC (n = 3)	0.76	–	
(b) Pit/orchestra source			
EDT (n = 1)	−0.80	−3.47	0.78 (s)
Δt_1 (n = 2)	−0.18	−0.21	29 (ms)
IACC (n = 3)	0.09	–	

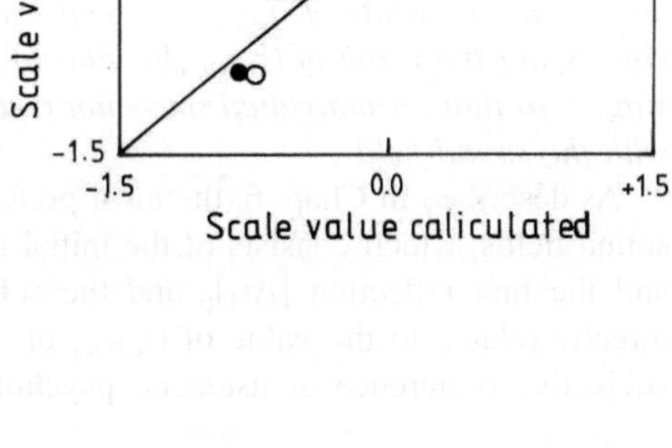

Fig. 9.9 Relationship between the scale value of balance obtained by PCTs and the scale values calculated by Eq. (9.2)

9.4 Conclusions

(1) EDT or the reverberation time was the major effects on the scale value of balance. And, effects of the spatial factor IACC was minor.
(2) The correlation coefficient between the measured and calculated scale values for the stage source is greater than that for the pit or orchestra source.
(3) The optimum values of the orthogonal factors are: $[\mathrm{EDT\ (stage)}]_p \sim 0.9$ s, ($[T_{sub}]_p \sim 1.1$ s), and $[\Delta t_1]_p \sim 27$ ms.
(4) Reverberation times (EDT) of both vocal and keyboard signals are almost the same as 0.82 s, or a slightly short value for the keyboard from the pit; the reverberation time of each existing opera house was apparently a little longer for the source position on the stage than that for the pit except one case in seven.
It is worth noticing that Sakai et al. (2000) reported that the averaged value of the preferred reverberation time T_{sub} of the vocal signal measured was 0.78 s with large individual differences 0.55–1.22 s, and the calculated value by the formulas with the value of $(\tau_e)_{min}$ is 0.53 s.
(5) The vocal sound source plays an important role in balance judgments.

9.5 Singing Styles on the Stage Blending with the Sound Field for Listeners

For musicians who wonder how their vocal sound on the stage blends with the sound field in a given opera house, it is because the sound field is regarded as "the second musical instrument." According to the theory of subjective preference, it should be deeply related to the value $(\tau_e)_{min}$ of the source signal. The temporal factors to be optimized are $[\Delta t_1]_p$ and $[T_{sub}]_p$ of the sound field for listeners, which are expressed in terms of $(\tau_e)_{min}$ as given by Eqs. (6.7) and (6.8), respectively. Here, we shall show how vocalists can control the value of $(\tau_e)_{min}$ changing performing style to blend with the sound field.

One method of controlling $(\tau_e)_{min}$ by a singer on the stage is to introduce "vibrato" and then "singing volume," which may play an important role in decreasing the value of $(\tau_e)_{min}$ to blend with short temporal factors. The amount of controlling the value of $(\tau_e)_{min}$ by vibrato is greatly dependent on each individual singer, so that an individual musician could acquire his/her own skill in blending with the sound field.

As described in Chap. 6, the most preferred conditions of the temporal factor of sound fields, which consists of the initial time delay gap between the direct sound and the first reflection $[\Delta t_1]_p$ and the subsequent reverberation time $[T_{sub}]_p$, are directly related to the value of $(\tau_e)_{min}$ of the running ACF of sound sources. The subjective preference of listeners' psychological response was well based on the

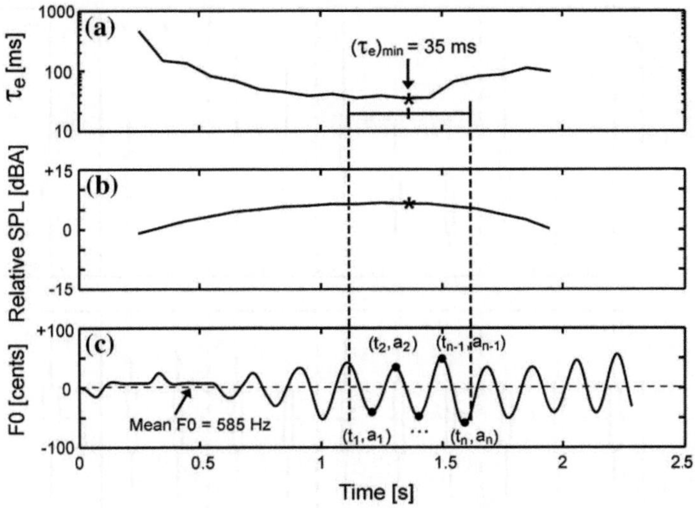

Fig. 9.10 Factors as a function of time analyzed for the vocal signal sung by a sopranist, "D5" with vowel/e/. **a** Value of τ_e extracted from the running ACF measured with 100 ms stepping interval ($2T$ = 500 (ms)). **b** Relative SPL measured with the A-weighting network and 100 ms stepping interval ($2T$ = 500 (ms)). **c** Vibrato waveform

brain activities (Chap. 4, Ando 2003), and it relates deeply to the aesthetic issue. Therefore, in order to blend a vocal source and the sound field in a given opera house, it has been attempted to understand how musicians can control the value of $(\tau_e)_{min}$ of the steady state piece of vowel signals by means of singing style. The singing style changes were (1) sound pressure level (SPL) as singing volume, (2) vibrato rate, and (3) vibrato extent (Kato et al. 2004). Figure 9.10 shows an example of the τ_e, the relative SPL and the pitch as a function of time, which were extracted from the running ACF. It is known that the most important piece of the signal is around the $(\tau_e)_{min}$, which is the most active part in the signal and deeply related to subjective preference judgments (Ando et al. 1989; Mouri et al. 2000).

Vibrato rate (VR) in the time domain and vibrato extent (VE) in the F_0 domain, respectively, are figured out, such that (Preme 1994, 1997)

$$\mathrm{VR}(k) = 1/[t_{k+1} - t_{k-1}],\ (\mathrm{Hz}) \tag{9.6}$$

$$\mathrm{VE}(k) = 1200\log_2[1 + |(a_{k-1} - 2a_k + a_{k+1})/(a_{k-1} + 2a_k + a_{k+1})|],\quad (\mathrm{cents}) \tag{9.7}$$

where k is the step in the time. Values of t_k and a_k (k = 1, 2, …, n) are found at positive and negative peaks of F_0 in the certain range of $(\tau_e)_{min}$. The mean values of VR and VE may be obtained in the certain time range.

Figure 9.11a shows the distribution of the $(\tau_e)_{min}$ values for sung vowels (10 vocal singers) for individual singer's sung with vowels. The $(\tau_e)_{min}$ values showed an

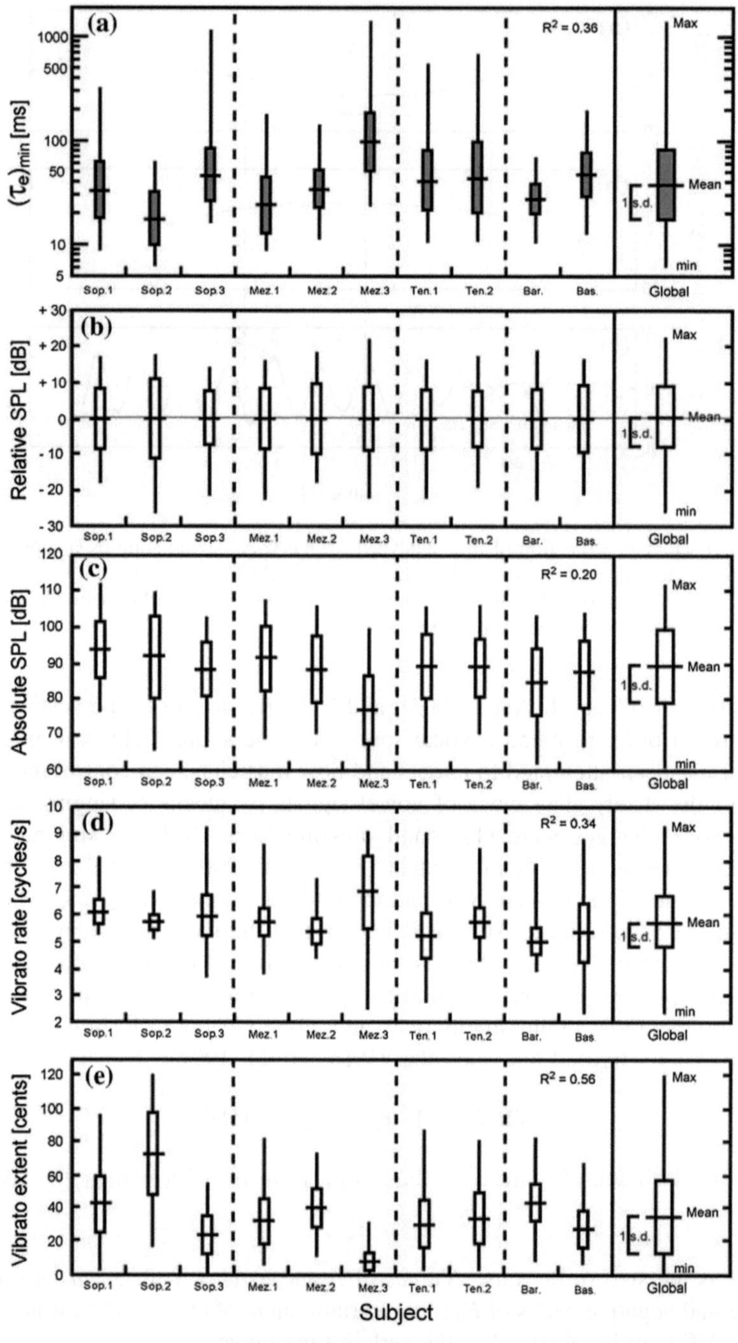

Fig. 9.11 Distribution of the factors for ten individual singers. **a** $(\tau_e)_{\min}$ value. **b** Relative SPL. **c** Absolute SPL. **d** Vibrato rate. **e** Vibrato extent

approximately normal distribution on a logarithmic scale, thus it is not a linear scale. The $(\tau_e)_{min}$ values for individual tones greatly and significantly ranged between 6.8 and 1,482 ms, and the geometric mean of individual singers ranged between 18 ms (Sop. 2) and 100 ms (Mez. 3). The geometric mean across singers was 39 ms, so that the most preferred reverberation time calculated by Eq. (6.8) is 0.9 s.

The relative SPL, absolute SPL, VR, and VE were dealt as explanatory variables for $(\tau_e)_{min}$ in this study. Figure 9.7b–e shows the distribution of these variables for individual singer's sung with vowels.

The standard deviation (SD) of relative SPL which shows dynamic range of SPL within each singer, varied between 7.5 dBA (Sop. 3) and 11 dBA (Sop. 2), while the global mean across singers was 9.0 dBA (Fig. 9.7b). Absolute SPL varied between 77 dBA (Mez. 3) and 95 dBA (Sop. 1), while the global mean across singers was 88 dBA (Fig. 9.7c).

Figure 9.7d shows observations of VR for individual tones ranged between 2.4 and 9.2 Hz, and the singer mean varied between 5.0 Hz (Bar. 1) and 6.9 Hz (Mez. 3). The global mean across singers was 5.7 Hz. The SD of each singer's Vibrato rate, which expresses intra-individual variation of Vibrato rate, varied between 0.3 Hz (Sop. 2) and 1.3 Hz (Mez. 3), and 0.6 Hz for global average across singers.

Figure 9.2e shows the results of VE for individual tones ranged between ±0.0 cents and ±120 cents and the singer mean varied between ±8.0 cents (Mez. 3) and ±73 cents (Sop. 2). The global mean across singers was ±36 cents. The SD of each singer's VE, which expresses intra-individual variation, varied between ±5.0 cents (Mez. 3) and ±24 cents (Sop. 2), and ±13 cents on global average across singers.

Table 9.3 indicates the correlation matrix among five quantitative variables. The value of $\log_{10}(\tau_e)_{min}$ significantly correlated with relative SPL (−0.37), absolute SPL (−0.52), and vibrato extent VE (0.72), $p < 0.01$. The $(\tau_e)_{min}$ value decreases with absolutely louder voice, and with greater VE. Due to the correlation coefficient, a method of controlling the $(\tau_e)_{min}$ value most effectively is mainly performed by VE as shown in Fig. 9.12.

In order to predict the absolute value of $(\tau_e)_{min}$, a linear prediction model that employs the main effects of the three factors and excluding individual differences can be formulated as

Table 9.3 Correlation matrix among five quantitative variables

	$\log_{10}(\tau_e)_{min}$	Relative SPL	Absolute SPL	Vibrato rate	Vibrato extent
$\log_{10}(\tau_e)_{min}$	1.00	**−0.37**	**−0.52**[1]	+0.34	**−0.72**[1]
Relative SPL		1.00	+0.89	+0.01	+0.22
Absolute SPL			1.00	−0.09	+0.41
Vibrato rate				1.00	−0.35
Vibrato extent					1.00

[1] $p < 0.01$

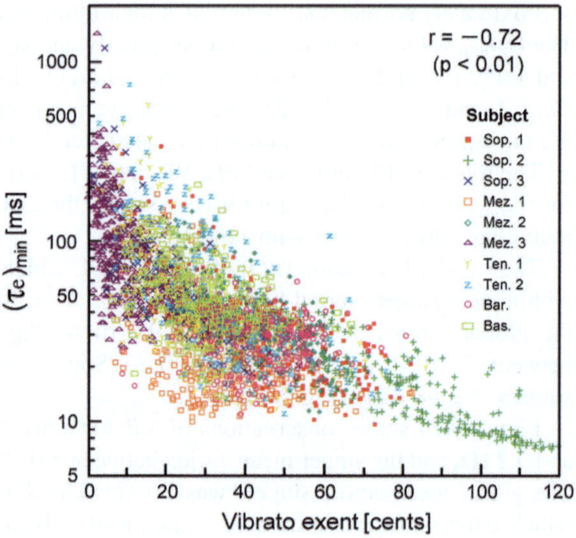

Fig. 9.12 Relationship between the $(\tau_e)_{min}$ value and VE for individual singer. *Different symbols* indicates results of vocal signals for each individual singer with all of vowels

$$\log_{10}(\tau_e)_{min} \sim a_1(\text{SV}) + a_2(\text{Pitch}) + a_3(\text{Vowel}) + k_1 \quad (9.8)$$

where a_1(SV), a_2(Pitch), a_3(Vowel), and k_1 are values calculated from a multiple regression analysis. Table 9.4 lists scores obtained in each category of a_1(SV), a_2(Pitch), and a_3(Vowel) by the analysis. The value of constant k_1 in Eq. (9.3) is equal to the logarithm of the geometric mean of $(\tau_e)_{min}$ across all singers. The results of coefficients or category scores for each condition and subject are shown in Fig. 9.13. According to Eq. (9.8) calculated values of $\log_{10}(\tau_e)_{min}$ and measured values are shown in Fig. 9.14.

Table 9.4 Category scores of each factor, a_1(SV), a_2(Pitch), and a_3(Vowel) in Eq. (9.3)

Factors	(Category)	Category scores
a_1(SV)	*pp (pianissimo)*	+0.17
	mf (mezzo forte)	−0.03
	ff (fortissimo)	−0.14
a_2(Pitch)	Low	+0.03
	Middle	−0.02
	High	−0.01
a_3(Vowel)	/a/	−0.09
	/e/	−0.04
	/i/	−0.01
	/o/	+0.02
	/u/	+0.12

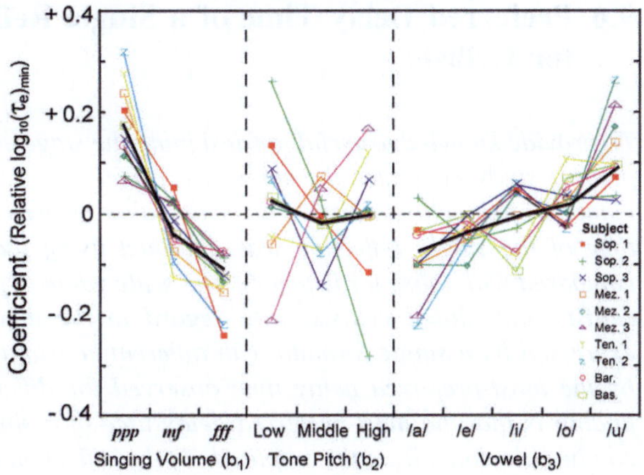

Fig. 9.13 Categorical score obtained as coefficient for each condition and subject

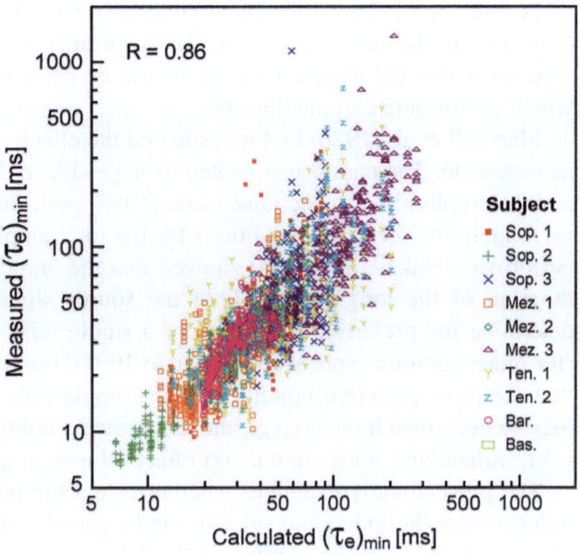

Fig. 9.14 Measured values of $(\tau_e)_{min}$ and calculated values by Eq. (9.2)

Consequently, we obtain

1. The $(\tau_e)_{min}$ value may decrease with greater vibrato extent most effectively, and with a louder voice as shown in Fig. 9.12 (see also Table 9.1).
2. The amount of contribution of the subjective singing volume to the value of $(\tau_e)_{min}$ (Fig. 9.13).
3. As shown in Figs. 9.12 and 9.14, large amounts of individual differences were observed, so that an individual musician could acquire his/her own skill in blending with the sound field.

9.6 Preferred Delay Time of a Single Reflection, Δt_1 for Cellists

To provide knowledge useful for designing the stage enclosure in a concert hall, the present study evaluates the subjective preference, with regard to ease of performance of five cello-soloists. The scale value of preference in change of the delay time of the single reflection was obtained using the PCT, and the results were compared with those for listeners. The scale value of preference for both individual cellists and global cellists, with regard to the delay time of reflection can be expressed by a single formula with different constants, normalizing the delay time by the most-preferred delay time observed for different music motifs. A notable finding is that the most-preferred delay time of a single reflection for each cellist can be calculated from the amplitude of the reflection and the minimum value of the effective duration $(\tau_e)_{min}$ *of the running ACF of the music signals as similar to the case of listeners for listeners (Chap. 6).*

In order to realize an excellent concert, we need to design the sound fields not only for the listeners but also in the stage area for the performers. The primary concern is that the stage enclosure should be designed to provide a sound field in which performers can play easily.

Marshall et al. (1978) first investigated the effects of stage size on the playing of an ensemble. The parameters related to stage size in their study were the delay time and the amplitude of reflections. Gade (1989) performed a laboratory experiment to investigate the preferred conditions for the total amplitude of the reflections of the performers. Nakayama (1984) showed that the amplitude of the reflection and the duration of the long-time ACF of the source signal in a similar manner could determine the preferred delay time of a single reflection for alto-recorder soloists (for more rigorous expression see: Ando 1998). Previously, it has been investigated that the most preferred condition of the single reflection for an individual singer may be described by the $(\tau_e)_{min}$ and a modified amplitude of reflection caused by the overestimate and bone conduction effect (Noson et al. 2000, 2002).

The present study examines whether or not the preferred delay time of a single reflection for the individual soloists can be calculated by the minimum value of the effective duration of the running ACF of the music signal played by the five cellists (Sato et al. 2000). Music motifs (motifs I and II) used in the experiments conducted by Nakayama (1984) were applied here. As shown in Fig. 9.15, the tempo of motif I was faster than that of motif II.

A microphone in front of the cellist picked up the music signal performed by each of five cellists. The distance between the microphone and the center of the cello body was 50 ± 1.0 cm. The music tempo was maintained with the help of a visual and silent metronome. Each music motif was played three times by each cellist. The minimum value of the effective duration $(\tau_e)_{min}$ of the running ACF of a music signal is the most active part of the music signal, containing important information and significant on the subjective preference. It was analyzed after passing through the A-weighted network with the integration interval, $2T = 2.0$ s,

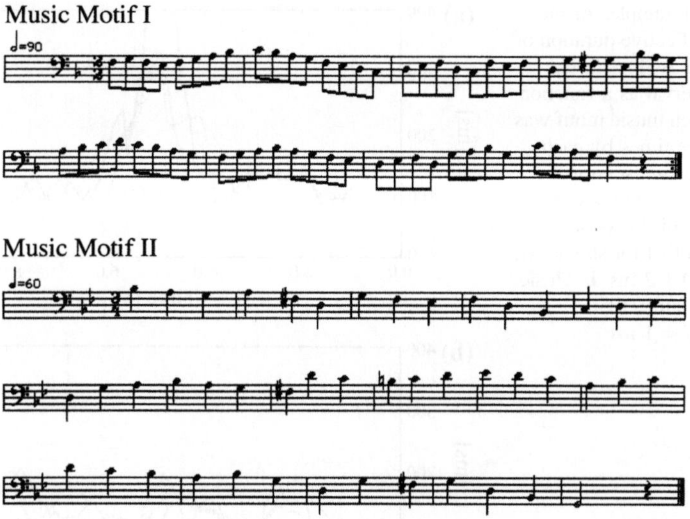

Fig. 9.15 Music scores of motifs I and II composed by Tsuneko Okamoto applied for experiment with Cellists (Ando 1998)

which was chosen according to Eq. (2.8). Usually, the envelope decay of the initial part of the ACF can be fitted by a straight line in the range from 0 to −5 dB to obtain the effective duration τ_e by the extrapolation at −10 dB as demonstrated in Fig. 2.5. Examples of effective durations of the running ACF for music motif I played by subjects B and E are shown in Fig. 9.16a, b. The minimum value of the effective duration $(\tau_e)_{min}$ of the running ACF for each cellist and each session are listed in Table 9.5. For all cellists, the effective durations $(\tau_e)_{min}$ for music motif I were about half of those for music motif II. Mean values of $(\tau_e)_{min}$ were 46 ms for music motif I and 84 ms for music motif II, and for both motifs the ranges of $(\tau_e)_{min}$ are within ±5 ms. Individual differences in the effective durations of the running ACF may depend on the performer's style.

The single reflection from the back wall in the stage enclosure was simulated in an anechoic chamber by a loudspeaker 80 ± 1.0 cm measured from the head of the cellist. The sound signal was picked up by a 1/2-inch condenser type microphone at the entrance of the performer's left ear and was reproduced by the loudspeaker after passing through a digital delay device. The amplitudes of reflection A_1, relative to that of the direct sound A_0 = 0 dB measured at the entrance of the performer's left ear was kept constant at −15 or −21 dB when the cellist played the musical note 'a' (442 Hz).

The preferred delay time of the single reflection was assumed to depend on the $(\tau_e)_{min}$ of the running ACF of source signal. The PCT was conducted for five sound fields, in which the delay time of reflection was adjusted for every cellist according to results of $(\tau_e)_{min}$ listed in Table 9.6. The subjects were asked which of the two sound fields was easier for them to perform in. The test consisted of 10 pairs ($N(N-1)/2$, N = 5) of stimuli in total, and for all subjects the test was

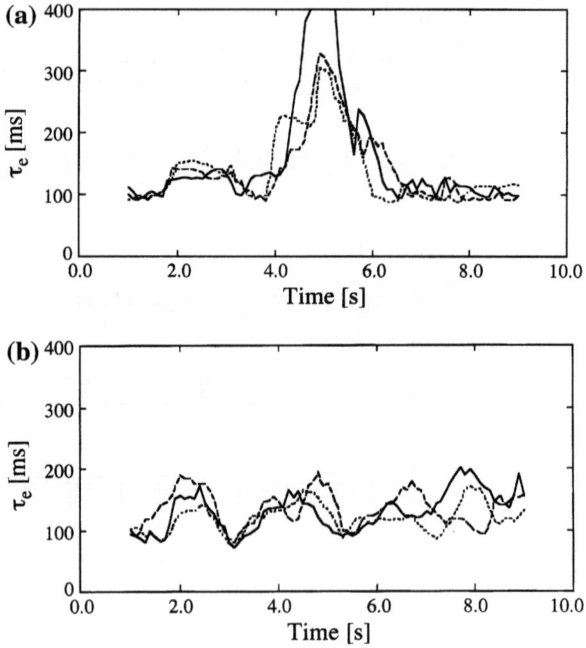

Fig. 9.16 Examples of the measured effective duration of the running ACF with the 100-ms interval as a function of time. Each music motif was played three times by each cellist. *Line* first session; *dashed line* second session; *dotted line* third session. **a** Music motif I for subject B, $(\tau_e)_{min}$ = 50 ± 2 ms. **b** Music motif I for subject E, $(\tau_e)_{min}$ = 37 ± 1 ms

Table 9.5 Minimum values of running τ_e of ACF for music motif played by each cellist

Cellist	Session	Music motif I (ms)	Music motif II (ms)
A	1st	35	90
	2nd	41	96
	3rd	41	89
B	1st	52	92
	2nd	49	87
	3rd	49	89
C	1st	37	89
	2nd	38	86
	3rd	36	93
D	1st	57	87
	2nd	56	85
	3rd	54	86
E	1st	37	71
	2nd	38	74
	3rd	36	79
Averaged		46	84

Table 9.6 Judged and calculated preferred delay times of a single reflection for cello soloists

A (dB)	A′ (dB) (=A + 10)	A′	Cellist	Judged $[\Delta t_1]_p$ (ms)		Calculated $[\Delta t_1]_p$ (ms)	
				Motif I	Motif II	Motif I	Motif II
−15	−5	0.56	A	16.2	47.9	16.3	38.5
			B	<12.0	73.8	35.2	62.7
			C	<12.0	60.8	21.3	51.3
			D	22.6	38.2	35.1	53.9
			E	17.6	63.6	17.3	35.2
			Global	18.0	48.3	24.3	47.5
−21	−11	0.28	A	18.1	48.4	21.8	51.5
			B	61.2	105.0	59.3	105.6
			C	–	77.9	–	80.6
			D	74.6	86.8	56.9	87.4
			E	<14.0	42.2	24.8	50.2
			Global	30.4	71.8	37.6	73.4

Calculated values of $[\Delta t_1]_p$ are obtained by Eq. (9.9) using the amplitude of the reflection A_1, and $(\tau_e)_{min}$ for music signal performed by each cellist

repeated thrice interchanging the order of the pairs. It took about 20 min for each cellist and for each music motif. Fifteen responses (5 subjects × 3 repeats) to each sound field were obtained and were confirmed by consistency tests. The scale values of preference for each cellist were obtained (Ando and Singh 1996; Ando 1998).

Figure 9.17 shows an example of the regression curve for the scale value of preference and the method of estimating the most preferred delay time $[\Delta t_1]_p$. The peak of this curve denotes the most-preferred delay time. The most-preferred delay times for individual cellists and the global preference results are listed in Table 9.6.

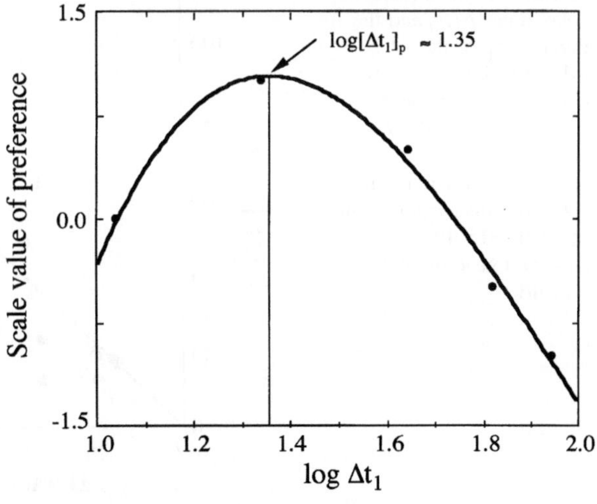

Fig. 9.17 An example of the regression curve for the preferred delay time (Subject D, Music motif I, −15 dB), $\log\,[\Delta t_1]_p \approx 1.35$, i.e., $[\Delta t_1]_p \approx 22.6$ (ms)

Global and individual results (except for that of subject E) for music motif II were longer than those for music motif I.

The most-preferred delay time of the single reflection also is described by the duration τ'_p of the ACF similar to that of listeners, which is expressed by

$$[\Delta t_1]_p = \tau'_p \approx [\log_{10}(1/k') - c'\log_{10}A'](\tau_e)_{min} \tag{9.9}$$

where the values k' and c' are constants that depend on a music instrument played. The value of A' is the amplitude of the reflection being defined by $A' = 1$ relative to −10 dB of the direct sound as measured at the ear's entrance. This is due to the overestimation of the reflection by a performer. This is called "missing reflection" of a performer.

Using the Quasi-Newton method, the values $k' \approx 1/2$ and $c' \approx 1$ are obtained. It is worth noting that the coefficients k' and c' for alto-recorder soloists were, respectively, 2/3 and 1/4 and for listeners it was, respectively, 0.1 and 1. After setting $k' = 1/2$, we obtained the coefficient c' for each individual as listed in Table 9.7. The average value of the coefficient c for the five cellists obtained was about 1.0. The relation between the most-preferred delay time $[\Delta t_1]_p$ obtained by preference judgment and the duration τ'_p of the ACF calculated by Eq. (9.9) using $(\tau_e)_{min}$ is shown in Fig. 9.18. Different symbols indicate the values obtained in different test series. The correlation coefficient between calculated values of $[\Delta t_1]_p$

Table 9.7 Coefficients c' in Eq. (9.9) for calculation of the preferred delay times of the reflection for individual and for global (the coefficient k' is fixed at 1/2)

	Cellist					Averaged (global)
	A	B	C	D	E	
Coefficient c'	0.47	1.61	1.10	1.30	0.67	≈1

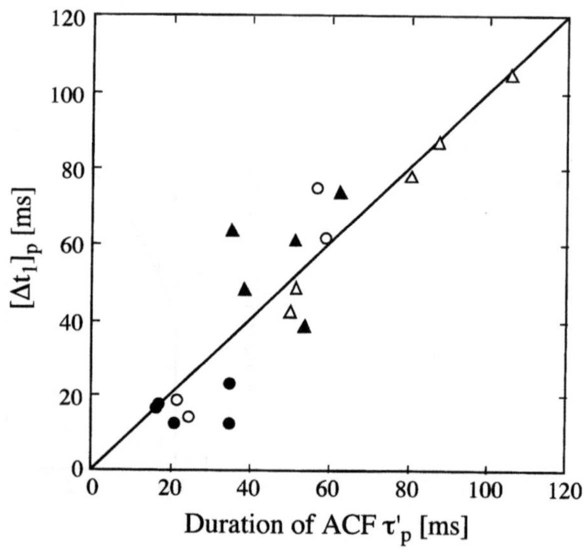

Fig. 9.18 Relationship between the most preferred delay time $[\Delta t_1]_p$ and the duration τ'_p of the ACF calculated by Eq. (9.8). Correlation coefficient, $r = 0.91$ ($p < 0.01$). *Filled circle* music motif I, −15 dB; *open circle* Music motif I, −21 dB; *filled triangle* music motif II, −15 dB; *open triangle* music motif II, −21 dB

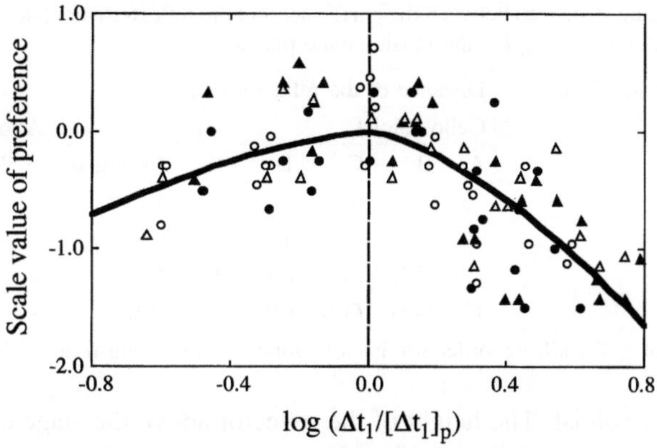

Fig. 9.19 Scale values of preference for each of five cellists as a function of the delay time of a single reflection normalized by its most preferred delay time calculated by Eq. (9.2). *Filled circle* Music motif I, −15 dB; *open circle* music motif I, −21 dB; *filled triangle* music motif II, −15 dB; *open triangle* music motif II, −21 dB. The regression curve is expressed by Eq. (9.2)

and measured values is 0.91 ($p < 0.01$). The scale values of preference for each of the five cellists as a function of the delay time of the single reflection normalized by the calculated $[\Delta t_1]_p$ are shown in Fig. 9.19. Different symbols indicate the scale values obtained in different test series. Each symbol has 25 data sets (5 subjects × 5 sound fields) except for the amplitude of −15 dB for music motif I (for which there were 20 data because consistency tests did not indicate a significant ability to discriminate preference in the results of Subject C). Although the scale values were obtained in different test series, tendencies are consistent with each other. The regression curve is expressed as

$$S = -\alpha|x|^{\beta} \tag{9.10}$$

where $x = \log \Delta t_1/[\Delta t_1]_p$, the power of x may be always fixed by $\beta = 3/2$ and the weighting coefficient α is 2.3 for $x \geq 0$ and 1.0 for $x < 0$.

The most-preferred delay time of a single reflection for each cellist can be calculated by Eq. (9.9) with the amplitude of the reflection and the minimum value of the effective duration $(\tau_e)_{\min}$ of the running ACF of the music motifs played by each cellist. The scale values of preference for both individual cellists and for global cellists with regard to the delay time of the single reflection can be expressed by such a simple formula, normalizing the delay time by the most-preferred delay time observed for different music motifs.

As an application, adjusting the height of the reflectors above the stage can control the delay time of a reflection. As listed in Table 9.8, the optimum distance between the performer and the reflector above the stage in relation to the minimum value of the effective duration $(\tau_e)_{\min}$ of the running ACF of the music program to be performed can be calculated. Here it is assumed that the distance between the instruments and the ear of the performer is 60 cm for a cello soloist and 20 cm for

Table 9.8 Optimum distances between the performer and the reflector calculated from Eq. (9.9) in relation to the value of $(\tau_e)_{min}$ for the music signal played

$(\tau_e)_{min}$ of the music signal (ms)	Distance of the reflector (m)						
	Cello soloist						Alto-recorder soloist
	A	B	C	D	E	Averaged	
30	3	10	6	8	4	6	2
50	6	21	13	16	8	13	4
70	9	(33)	21	(26)	13	20	6
90	13	(46)	(30)	(36)	18	(29)	8

Note The value of τ_e for alto-recorder soloist was obtained for a long-time ACF ($2T$ = 32 s)

an alto-recorder soloist. The height of the reflector above the stage can be adjusted if the minimum value of the effective duration $(\tau_e)_{min}$ of the running ACF of the music to be played is measured before performance. For practical convenience, this adjustment may be made in the real sound field with the reverberation. Note that in this situation, the total amplitude of the reflections might replace the amplitude of the single reflection similar to the case of listeners.

In concluding this section, Fig. 9.20 shows the relative amplitude of a single reflection to that of the direct sound for the preference of cello-soloists as a function of the delay time of a single reflection normalized by the minimum value of the effective duration $(\tau_e)_{min}$ of the running ACF, as well as several other subjective responses as a function of the delay time of a single reflection normalized by the value of the effective duration τ_e of the long-time ACF of the source signal. All these values can be calculated by Eq. (9.9) with constants k and c for each subjective response. The alto-recorder soloist's preference is also plotted in this figure. The values for cellists are close to the threshold of perception (aWs) for listeners. These reconfirm the phenomenon of "missing reflection" for performers.

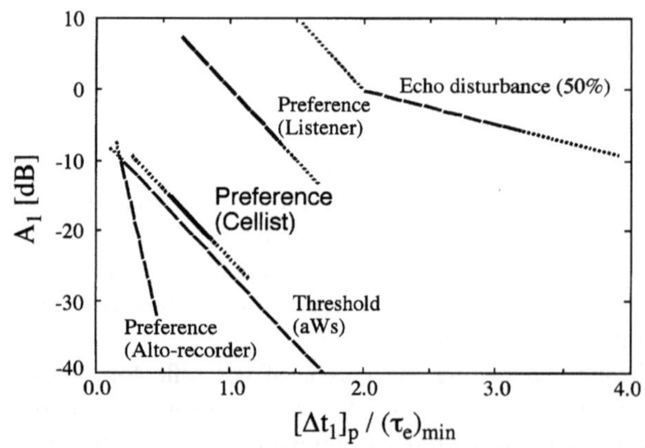

Fig. 9.20 Relative amplitude of the single reflection for the preference of cello-soloists as a function of the delay time of a single reflection normalized by the value of $(\tau_e)_{min}$. For additional information, the amplitudes of other subjective responses as a function of the delay time of the single reflection normalized by the value of $(\tau_e)_{min}$ of the source signal are also plotted (Ando 1998)

Chapter 10
Optimizing Room-Forms

The theory of subjective preference in terms of the four orthogonal acoustical factors has been described in Sect. 6.3 (Ando 1985, 1998) allows us to evaluate at each seat of the sound field in an enclosure under design. The linear scale value of subjective preference has been obtained by using the law of comparative judgment. The units of the scale value derived from a series of experiments with different sound sources and different subjects were almost constant, so the scale values may be added as expressed by Eq. (6.9). Here, an application of the GA to the system for optimizing the shape of an enclosure is available with the spatial factor (IACC) of sound fields, which is associated with the right hemisphere. The first model was an optimization of the proportions of a hall of the typical shoebox type. The results show that the optimized form is similar to the "Grosser Musikvereinsaal" in Wienna. The second model is the optimization of the shape with a number of portions of modification. A kind of the leaf-shaped plan is a typical result of the maximization of the scale values of subjective preference in relation to the IACC for the audience area.

10.1 Genetic Algorithm for Optimal Shape-Design

The temporal and spatial factors are carefully designed, in order to satisfy both left and right human cerebral hemispheres for each listener, respectively. The GA (Holland 1975), a form of evolutionary computing, has been applied to a variety of complex engineering problems. The algorithm is started with a set of solutions (represented by 'chromosomes') that is called a population shown in Fig. 10.1. Solutions from one population are selected to form a new population. This is repeated until a condition (for example, an improvement over a previous best solution) is satisfied.

In this study, a GA system was applied to the design of enclosures (Sato et al. 2002, 2004). The GA system was used to generate the alternative scheme. Those architectural schemes, which produce greater scale values of subjective preference, are selected in the process of evolution. We started by applying this technique to

Y. Ando, *Opera House Acoustics Based on Subjective Preference Theory*,
Mathematics for Industry 12, DOI 10.1007/978-4-431-55423-3_10

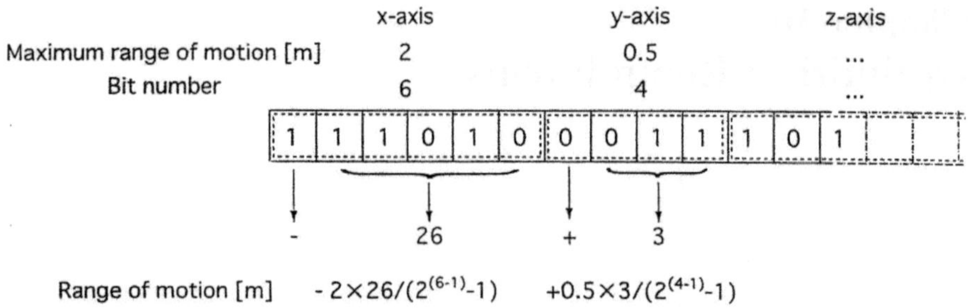

Fig. 10.1 An example of the binary strings used in encoding of the chromosome to represent modifications to the room shape

optimize the proportions of the most basic form that is a shoebox shape of enclosure. The scale value of subjective preference is employed as fitness function. Those enclosure shapes that produced greater scale values are selected as parent chromosomes. To create a new generation, the room shapes are modified and the corresponding movement of the vertices of the walls is encoded in chromosomes, i.e., binary strings. After GA operations that included crossover and mutation, new offspring were created. The fitness of the offspring was then evaluated in terms of the scale value of subjective preference. This process here was repeated until the end condition of about 2000 generations had been satisfied.

The typical spatial factor is IACC for a source on the stage were calculated at each of a set of seats. The single omnidirectional source is located at the center of the stage, 1.5 m above the stage floor. The receiving points that correspond to the ear positions were 1.1 m above the floor of the hall. The image method is applied to determine the amplitudes, delay times, and directions of arrival of reflections at these receiving points. Reflections were calculated up to the second order to reduce the calculation time. Note that second-order reflection is enough to provide convergence of the physical factors for a listening position near the stage. The averaged values of the IACC for five music motifs (Motifs A through E, Ando 1985) were applied.

According to architectural scheme under design, the scale value in relation to each orthogonal factor S_i (i = 1, 2, 3, 4) can be calculated by Eq. (6.10). Here, the parameters x_i and coefficients α_i are listed in Table 6.2. In this calculation, the scale values of subjective preference due to IACC, i.e., S_4, are applied as the measure for the sake of simplicity, because the geometrical shape of a hall directly affects significantly this spatial factor. Before going into GA operation, it is highly recommended to have a good initial shape of room so as to be small values of IACC due to recommended angles centered on 55° left and right sides from the media plan (Sect. 6.3.4) for early reflections to each seating position that is important to obtain a final scheme of a opera house without a lot of time or calculations.

Especially, the scale values of S_2 and S_3 were excluded, because due to the temporal factors that are much related to the value of $(\tau_e)_{min}$. Changing the performance style of vocal sound on the stage and music in the pit (Chap. 9), these temporal factors may be satisfied. It has been found that music and vocal sound with rapid movements sounds or vibrato are best fitted in an opera house with a short initial time delay gap Δt_1, and a short subsequent reverberation time T_{sub}. A slow tempo music performed in the pit is blended by relatively long values for the factor Δt_1 and T_{sub} that is classified by the value of $(\tau_e)_{min}$.

An example of the encoding of the chromosome is given in Fig. 10.1. The first bit indicates the direction of motion for the vertex. The other $n - 1$ bits indicate the range over which the vertex is moved. Here, a rather simple room shape is recommended to reduce the calculation time, and the single binary string has 140 bits at most. However, it is possible to process the binary string of 300 or 400 bits, if we had more time for calculation at design stage of opera house.

In the next crossover step, genes are selected from parent chromosomes and used to create a new offspring. Some crossover point within chromosome is chosen at random and everything before this point is copied from the first parent while everything after this point is copied from the second parent. After the process of such a crossover, mutation is applied. This is to prevent all solutions in a population falling into a locally optimal solution to the problem. Mutation is the application of a random change to the new offspring. A few randomly chosen bits of the chromosome are switched from 1 to 0 or from 0 to 1.

10.2 A Simple Example of Designing a Shoebox-Type Room

First of all, the proportion of the shoebox hall has been optimized. The initial geometry was the hall with 20-m wide, the stage of 12-m deep, 30-m long, and the ceiling of 15 m above the floor. The point source was located at the center of the stage and 4.0 m from the front of the stage and 72 listening positions were selected. The range adjusting each sidewall and the ceiling is ±5 m from the respective initial positions, and the distance through which each was moved, was coded on the chromosome of the GA. The scale value at the listening positions other than those within 1 m of the sidewalls were included in the average $(\overline{S_4})$. As is well known that in all subjects tested the preference increases with decreasing IACC (Ando 1985; Singh et al. 1994) commonly.

The result of GA operation is indicated in Table 10.1. It is similar to the proportions of the "Grosser Musikvereinsaal," which was said the most excellent concert hall in that time. The length/width ratios resulted are almost the same as those of "Grosser Musikvereinsaal" (Fig. 10.2).

Table 10.1 Comparison of proportions for the optimized spatial form by S_4 (IACC) of "shoebox type," and the "Grosser Musikvereinsaal"

	Length/width	Height/width
Optimized for $\overline{S_4}$ for the IACC	2.57	1.43
Grosser Musikvereinsaal	2.55	0.93

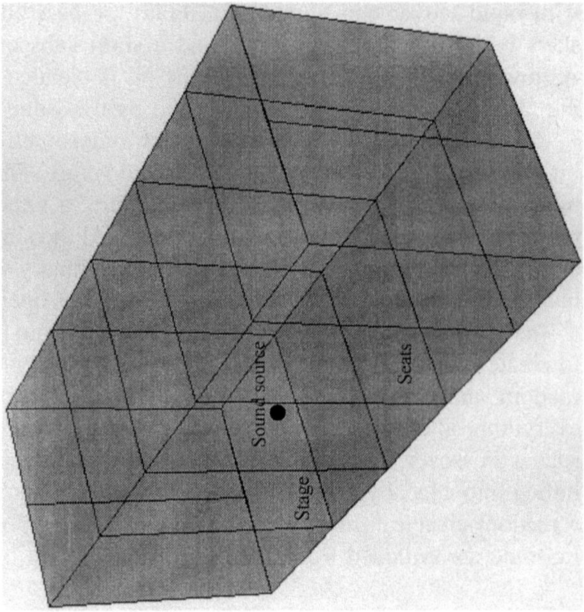

Fig. 10.2 The initial scheme of a concert hall. The range of sidewall and ceiling variation was ±5 m from the initial scheme

10.3 A Shape Improved from the Shoebox-Type Room

In order to attain knowledge of opera house design from a view point of concert hall, the floor plan optimized above result is selected as a starting point.

10.3.1 A Shape Improved from the Shoebox-Type Room

According to the above results, an initial form was 14-m wide, the stage was 9-m deep, the room was 27-m long, and the ceiling was 15 m above the stage floor. The sound source was again 4.0 m from the front of the stage, but was 0.5 m to one side of the centerline and 1.5 m above the stage floor. The front and rear walls were vertically bisected to obtain two faces, and each stretch wall along the side of the seating area was divided into four faces. The walls were kept vertical (i.e., tilting was not allowed) to examine only the plan of the hall in terms of maximizing $\overline{S_4}$. Each wall was moved independently from other walls. In the acoustical simulation using image method, the openings between walls were assumed not to reflect the sound. Forty-nine listening positions distributing throughout the seating area on a 2×4 m grid were selected. In the GA operation, even though the sidewalls were moved, these 49 listening positions were all included. The moving range of each vertex was ±2 m in the direction of the line normal to the surface. The coordinates of the two bottom vertices of each surface were encoded on the chromosomes for

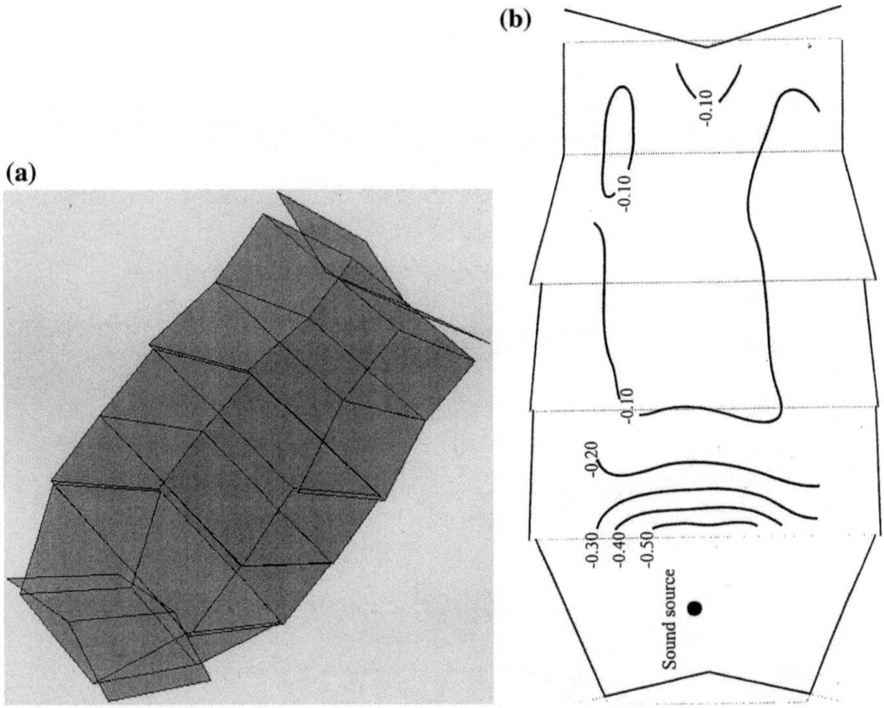

Fig. 10.3 **a** Resulting shape of the concert hall optimized by the typical spatial, IACC. **b** Contour lines of equal $\overline{S_4}$

the GA. In this calculation, the most preferred listening level was assumed for a point on the hall's long axis (central line), 10 m from the source position for the sake of convenience.

The result of optimizing for $\overline{S_4}$ is shown in Fig. 10.3a and contour lines of equal $\overline{S_4}$ values are shown in Fig. 10.3b. To maximize $\overline{S_4}$, the rear wall of the stage and the rear wall of the audience area took on convex shapes that avoid reflections from near the median plane.

10.3.2 Actual Design of a Leaf-Shape Room

An example of applying this design theory was performed in the Kirishima International Concert Hall as a leaf-type, in cooperation with the architect Fumihiko Maki in 1992 as shown in Fig. 10.4a–c (Maki 1997; Ando et al. 1997; Nakajima and Ando 1997). Acoustic design elements were as follows (Ando 1998, 2007):

(1) A leaf-shape plan was applied, (2) the sidewalls were tilted, and (3) the ceiling consisted of triangular plates.

These have realized a small value of the IACC at nearly every seat.

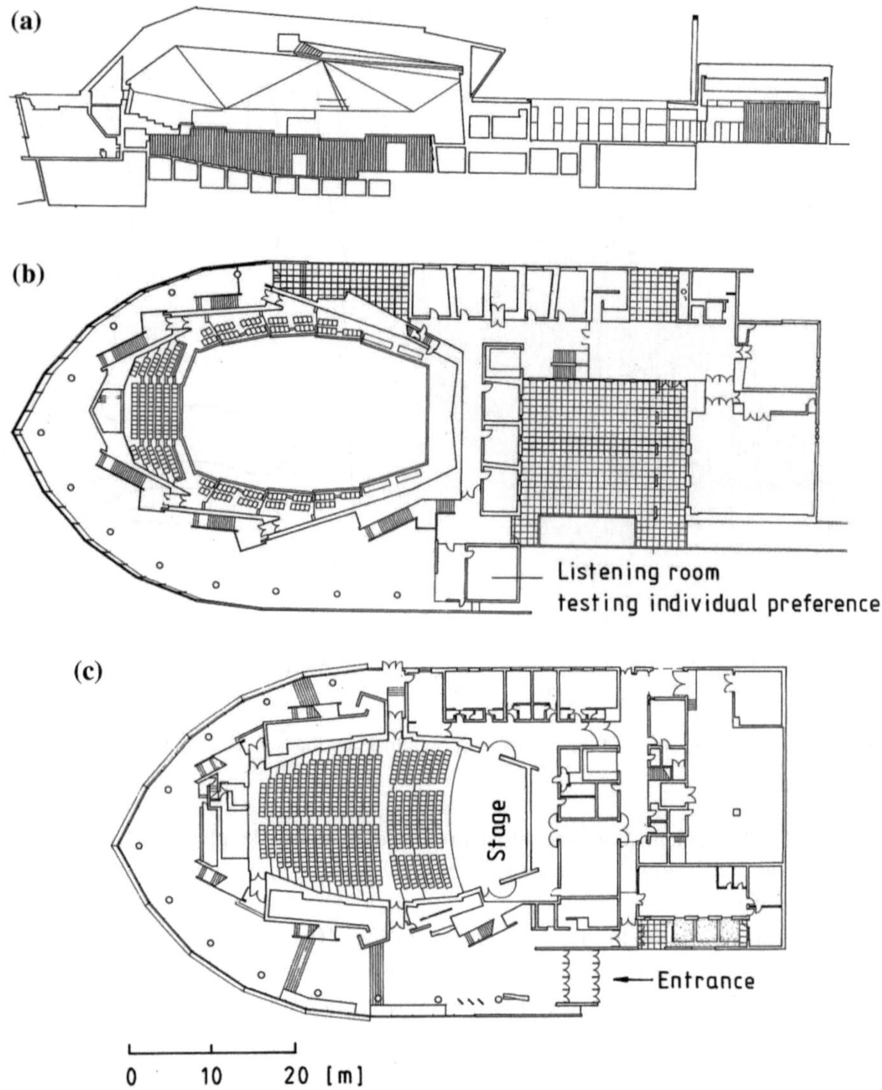

Fig. 10.4 The Kirishma International Music Hall designed and built in 1994

10.4 Effects of Scattered Reflection of a Canopy Array

In additional to optimization of room shape, obtaining a knowledge of scattered reflections of a canopy array for decreasing IACC in the area of audience floor of a theater and an opera house are discussed. If hight of canopy is changed, then according to $(\tau_e)_{min}$ *of the sound sources to be performed on the stage and in the pit can be well blended by the sound field of an opera house.*

10.4.1 Transfer Function for Panel Arrays

The arrays shown in the right column of Fig. 10.5a–c composing of the three different shapes of reflectors, are examined. Each array has 35 panels, the total area of the array is 280 m^2, and the total panel area is 140 m^2, so that the ratio of these areas is 50 %. The transfer functions shown in these figures are calculated when the sound wave impinges at the center of the array with the incident angle $\theta = 45°$. In the left column of Fig. 10.6, the solid and dotted curves represent the calculated results for the panel arrays and for single panels, respectively.

Obviously, the large dips in the transfer function of a triangular panel array in the low frequency range are much smaller than that of the others. When a geometrical ray reflection exists on a single panel, the transfer function in the high frequency range is almost same as that of the central single panel in the array. There are remarkable low frequency components which do not exist for the reflection of a single panel. This phenomenon is caused by diffraction effects of neighboring multiple panels, as demonstrated (Sect. 10.4.2).

For further information, the solid lines in the figure indicate Rindel's estimation lines of the transfer function for a rectangular panel array (Rindel 1986). The

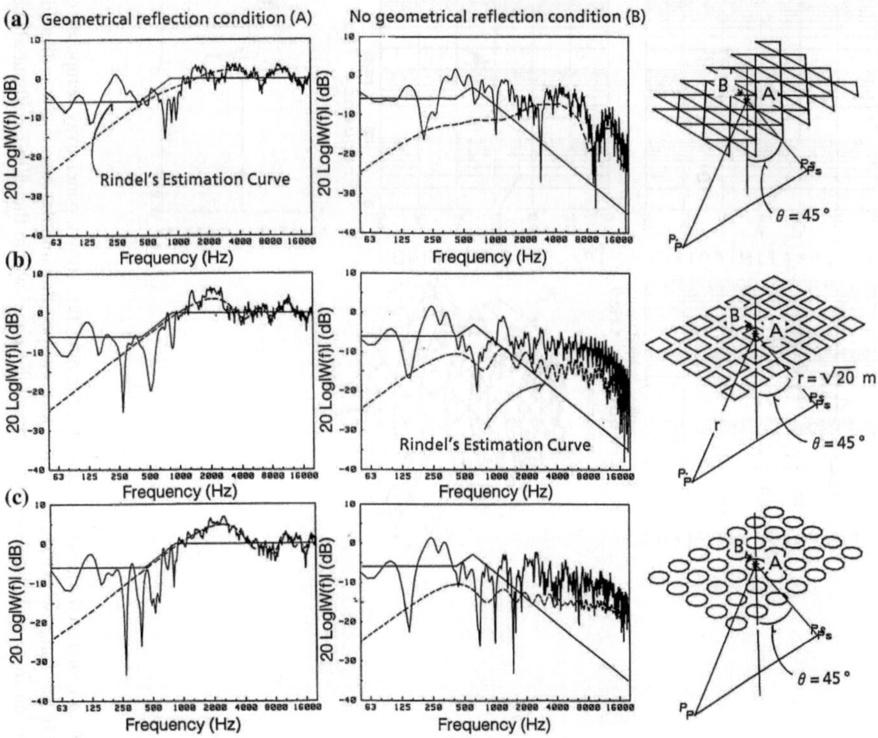

Fig. 10.5 Calculated transfer function for the reflecting arrays from canopy. **a** Array of *triangle* reflectors. **b** Array of *square* reflectors. **c** Array of *decagon*

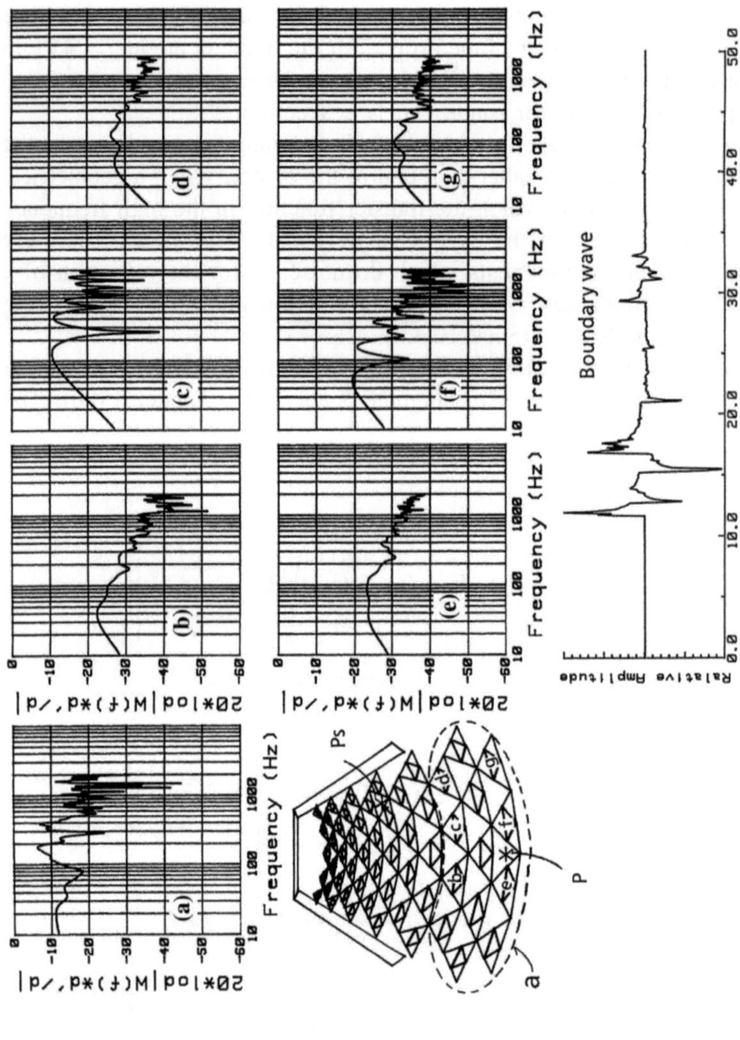

Fig. 10.6 **a** Calculated transfer function for the reflection from a panel array composed of the 13 nonplanar panels within ellipse indicated by *a*, which were installed in the Tanglewood Music Shed. The corresponding impulse response is indicated on the *lower part* of this figure. **b–g** Calculated transfer function for the reflection from each single nonplate triangular panel within the ellipse without all neighboring panels (Nakajima et al. 1992)

amplitude of transfer functions for panel arrays are in close agreement with Rindel's estimation only in the case when the path of geometrical reflection exists on the center of the panel.

10.4.2 Lateral Reflection Components from Overhead Canopies

In the Boston Symphony Orchestra's Tanglewood Music Shed, the canopy, which consists of nonplanar triangular panels, plays an important role in decrease the IACC, since there are no side wall reflections due to the wide fan shape of the Shed. The sizes of the triangular canopy panels range from 2.5 to 8.0 m, and the opening area are triangular as well and of the same dimensions as the panels (50 %). The canopy is suspended about 6.5 m above the audience floor and extends over the stage as well as the front part of the audience area. Figure 10.6a is a typical example of the transfer function calculated for the panel array composed of 13 nonplanar triangular panels located within the ellipse drawn in the figure. The related impulse response is shown at the bottom of the figure. Figure 10.6b–g shows the transfer function for particular receiving points. In these figures, 0 dB refers to the level of the direct sound from the source to a receiving point without any reflection. It is remarkable that relatively strong low frequency components arrive from panels away from the median plane, as demonstrated in the transfer functions. These reflections are adequate to decrease IACC for the audio frequency range. The high frequency components from the panel directly in line above, help to avoid the image shift of the sound source, keeping the maximum value of the interaural crosscorrelation function at the time origin, $\tau_{IACC} = 0$.

In another example of an existing hall, designed by Nakajima et al. (1992), triangular reflectors are installed only above stage. Triangular reflectors with a angle of about 120° show effective reflections for a wide frequency range. When such reflectors are installed above stage, then the lateral reflections in the low frequency range may serve to decrease the IACC.

Results of measured IACCs with the triangular reflectors above the stage only are shown in Fig. 10.7a and calculated IACC without the reflectors are shown in Fig. 10.7b. When the reflectors are installed above stage, then the IACC values of seats close to the stage are decreased. According to the effective duration of ACF of program sources, this kind of reflector above stage is quite useful for musicians as well. This supports their preferred performance by controlling the delay time of reflections from the height of canopy according to the value of $(\tau_e)_{min}$ of the source signal.

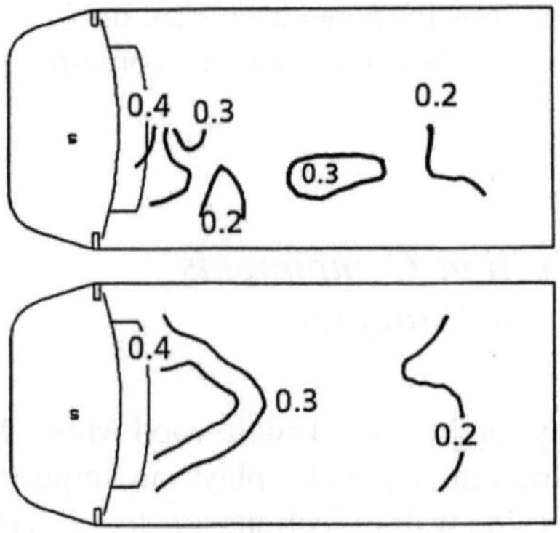

Fig. 10.7 The IACC with music signal (Sinfornietta, Opus 48; IV movement composed b Malcolm Arnold) in the existing concert hall with canopy above the stage similar to Fig. 10.6. *Above* Measured values with panel array composed of the seven nonplanar triangular panels. *Below* Calculated value without any canopy

10.5 Acoustic Design Proposal for an Opera House

Two different kinds of sound sources in an opera house, i.e., the vocal source on the stage with a relatively short value of $(\tau_e)_{min}$, *and the orchestra music in the pit with a long value of* $(\tau_e)_{min}$. *For these two quite different source signals, a proposal of designing an opera house using an acoustically transparent floor is made here based on the theory of subjective preference.*

The theory of subjective preference has been reconfirmed testing sound fields in an existing opera house. In an opera house, the temporal factors (Δt_1 and T_{sub}) for two different source signals should be carefully designed as indicated in Table 10.2, as well as the spatial factor (LL and IACC) for both of two sources.

Table 10.2 A typical example of temporal and spatial factors to be optimized for acoustic design of an opera house

Source location and the value of $(\tau_e)_{min}$	Temporal factor Δt_1	Temporal factor T_{sub} (s)	Spatial factor LL (dB)	Spatial factor IACC
Stage (vocal) $(\tau_e)_{min} \approx 20$ ms[a]	≈20 ms	≈0.5	<3.0	<0.5
	≈(A = 1.0)			
Orchestra music in the pit $(\tau_e)_{min} \approx 40$ ms[b]	≈20 ms	≈1.0	<3.0	<0.5
	≈(A = 3.0)			

[a] The mean value for different vowels and pitches (Kato et al. 2004)
[b] A possible minimum value of the orchestra music (Ando 1998)

10.5.1 Considerations Due to the Temporal Factor

The acoustic design of theaters has been made only the space above floor, except for the ancient Greek theaters (Vitruvius, ca. 25 B.C.). Since the acoustic field below the audiences' ears is equally important as one above the ears, we may take the under-floor space into consideration in designing sound fields.

(1) By utilization of the under-floor space in addition to the above-floor space, we may control the temporal factor of two different source signals, for example, the orchestra music with the value of $(\tau_e)_{min} > 40$ ms in the pit and the vocal sound with $(\tau_e)_{min} \approx 20$ ms on the stage (Table 10.2). The most important acoustic design of an opera house is made for the vocal sound source on the stage. For this purpose, the upper space of audience area should have the short initial time delay of early reflection $(\Delta t_1)_{vocal}$ and the reverberation time $(T_{sub})_{vocal} \approx 0.5$ s at each seat.
(2) For the orchestra music in the pit, a well-designed acoustically transparent floor below auditorium including the orchestra pit realizes a large space, so that a preferred reverberation time $(T_{sub})_{music} \approx 1.0$ s particularly for a low frequency range due to Eq. (6.8) can be realized.
(3) In addition, it is known that there is the SPL-dip in low frequency in the seating area that is caused by the interference effect of the direct sound and the reflected sound of the floor in the audience area. To eliminate the SPL-dip in the low frequency, as a matter of fact, it has been realized by utilizing the under-floor space with acoustically transparent floor (Takatsu et al. 2000). In the frontal area close to the stage, 5 mm diameter holes have been drilled through to the under-floor space in a 15 mm × 15 mm grid. A part of floor under the chair legs, there were drilled holes of a 25 % ratio to the extent of strength permits. This allows sound wave to pass through to the under-floor space eliminating the dip of low frequency range caused by the interference effect.
(4) For vocal sound on the stage $(\Delta t_1)_{vocal} \approx 20$ ms obtaining the preferred initial time delay of reflection according to Eq. (6.7) where $A_1 = A$ in the frontal seating area with the total amplitude of reflections $A = 1.0$, a canopy comprising triangular plates may be installed. At the same time, this may play important role providing enough sound energy needed between the vocalist on the stage and the performers in the pit. If the height of the canopy can be adjusted, then proper $(\Delta t_1)_{vocal}$ may be kept according to the value of $(\tau_e)_{min}$ of different styles of vocal signals.

10.5.2 Considerations Due to the Spatial Factor

(1) It has been shown that such a canopy above the pit play important role to decrease the IACC for audience area also (Nakajima et al. 1992).

(2) In order to obtain a small value of the IACC for two different sound sources at audience floor close to stage, for example, a leaf shape of the plan can be applied as realized in the Kirishima International Concert Hall (Ando 1998). The side walls may supply enough energy of early reflections for listeners arriving from centered on ±55° measured from their median plane.
(3) Another important factor is the balance of LL for listeners from both the singer on the stage and orchestra in the pit (Sect. 9.3).

10.5.3 Acoustic Design Proposal for an Opera House

As shown in Fig. 10.9, we consider two different spaces for the temporal acoustic design. Supposing a certain amount of transmission loss of the floor and an absorption of audience for the mid and high frequency components of vocal sound on the stage, the temporal design in the space above the floor is made. For orchestra sound in the pit including low frequency components, on the other hand, the transparent floor with less transmission loss connects spaces under and above the floor to be one acoustically large space.

A proposed scheme of opera house is shown in Figs. 10.8, 10.9, and 10.10. In this plan of opera house (Fig. 10.8), frontal panels of boxes form a leaf-shape as similar to the Kirishima International Concert Hall supply the useful reflections. This kind of shapes realizes to make decrease the IACC at each seating position of audience floor controlling its angles for early reflections to listeners (Sect. 10.3.2; Ando 1985, 1998). As shown in Figs. 10.8, 10.9, and 10.10, the canopy array above frontal areas of stage and a reflector in front of pit can control the balance of the LL

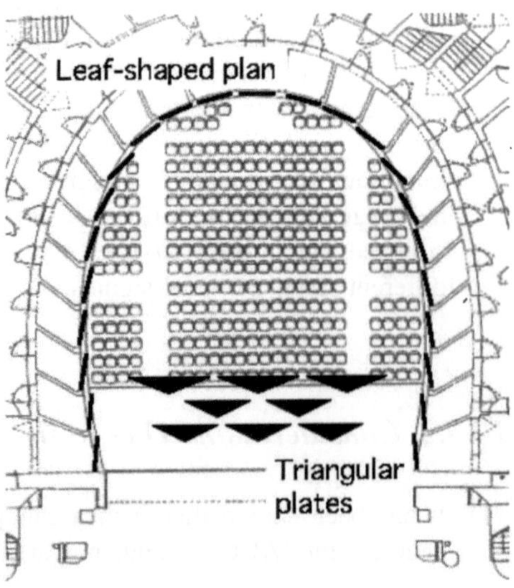

Fig. 10.8 Plan of proposed opera house with the leaf-type shape

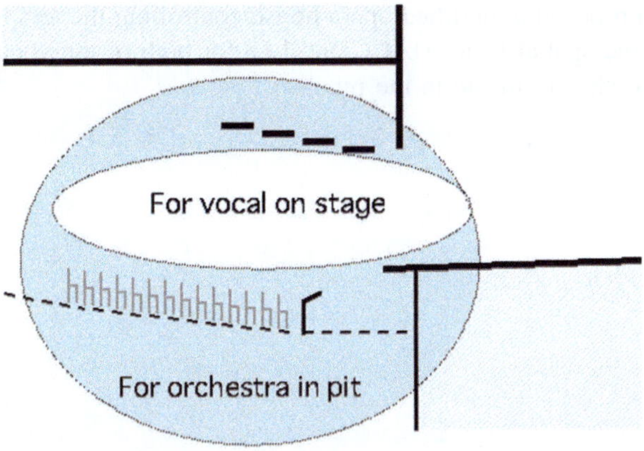

Fig. 10.9 An acoustic design proposal for the two different sound sources, the vocal on the stage and orchestra in the pit. **a, b** Cross-sections. Transparent floor together with *bottom shape* and canopy in an opera house for both performer and audience close to the stage decrease the values of IACC. Note that the values of IACC in the seating areas close to stage usually indicate large values due to the strong direct sound

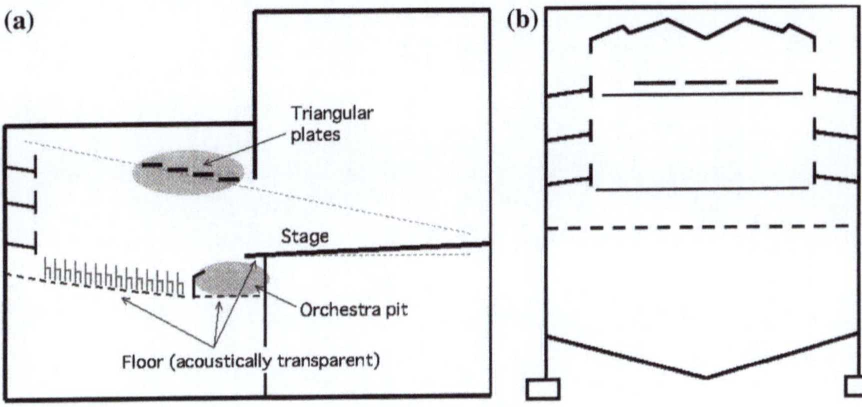

Fig. 10.10 Proposal of large space under floor with transparent sound for lower frequency range avoiding a large dip due to interference effects by the direct sound and reflection from the hard floor

of the vocal sound and the orchestra sound for audience. This may produce the relatively short initial delay time of reflection, $(\Delta t_1)_{\mathrm{vocal}}$, at the same time. These may provide enough sound energy between musicians in the pit to the stage performers realizing the communication in performance for ensamble. Figures 10.8, 10.9 and 10.10 demonstrate one large acoustic space with the transparent floor for the orchestra music in the pit obtaining the relatively long reverberation time, $(T_{\mathrm{sub}})_{\mathrm{music}}$. The shape of bottom with deeper center part in the opera house also may act to reduce the IACC.

So far, we proposed a modified opera house controlling the temporal factor (Δt_1 and T_{sub}) and the spatial factor (IACC and LL) for both of the vocal sound on the stage and the orchestra music in the pit.

Chapter 11
Visual Sensations on the Stage Blending with Opera and Music

In visual design on the stage, temporal and spatial aspects of the visual field could be taken into consideration in blending with opera and music, so that the dead stage lives. To attain the fundamental knowledge of preferred conditions of vision, subjective preference tests were conducted by changing temporal factors and spatial factors, which are extracted from the temporal ACF of target signals and the spatial ACF of the visual field, respectively. Results of the most preferred condition of flickering light, movements of a single target, and two-dimensional textures are explicitly described by the respective factors. The visual scene on the stage of opera house is well designed according music and story to be performed.

11.1 Visual Pitch Perception of Complex Signals

It has been found that perceived "pitch" at the fundamental frequency of visual complex signals, even in the random-phase conditions, in which the period of the fundamental is unclear in its real waveform. One promising operation to extract such a periodicity in the visual signal is the factor τ_1 extracted from the ACF of target signals.

This section describes a phenomenon that is analogous to the auditory–brain system. It is called "missing fundamental," which is known in the auditory pitch sensation (Sect. 5.1.1). Previously, some studies in vision were related to compound waveforms (de Lange 1954; Bowen et al. 1989; Bowen et al. 1992; Kremers et al. 1993; Eisner 1995), in which square and saw-tooth waveforms were commonly applied in comparison with sinusoidal waves. Each square and saw-tooth waveforms consists of the fundamental frequency (F_0) and a series of sinusoidal components (harmonics). However, no studies in the temporal vision that dealt with a compound waveform without the F_0 component have been performed.

The effect of F_0 is known in the spatial vision only (Henning et al. 1975; Nachmias and Rogowitz 1983). Henning et al. (1975) reported that in their experiment on the simultaneous masking of vision, the F_0 component not being contained in the masking stimulus affected the detection of the test stimulus.

Y. Ando, *Opera House Acoustics Based on Subjective Preference Theory*,
Mathematics for Industry 12, DOI 10.1007/978-4-431-55423-3_11

Four subjects, males aged 23–26-years old, participated in this experiment. All had normal or corrected-to-normal vision. They dark-adapted for about 1 min before all sessions. The light source was a 7-mm-diam green light-emitting diode (LED) set at a distance of 0.8 m from the observer in dark surroundings. The LED stimulus field was spatially uniform and its size corresponded to 0.5°

Stimuli in the present study were compound waveforms consisting of five complex components without fundamental frequency. The frequency of each component corresponded to the *n*-th harmonic of the fundamental frequency F_0. In Series A, we selected four stimuli in terms of the complex frequency range with $F_0 = 1$ Hz. Stimulus 1 consisted of 3, 4, 5, 6, and 7 Hz, and for stimuli 2, 3, and 4 the frequency ranges were selected, respectively, 11–15, 21–25, and 31–35 Hz. In Series B, for stimuli 5, 6, 7, and 8, complex components were selected in the frequency range for 30–40 Hz, in which we cannot detect any flickering rate if only a single component is presented. Stimulus 5 with $F_0 = 0.75$ Hz consisted of 30, 30.75, 31.5, 32.25, and 33 Hz. For the stimuli 6, 7, and 8 (with $F_0 = 2$, 2.5, and 3 Hz, respectively), the components were 30–38, 30–40, and 27–39 Hz, respectively.

The waveforms of the complex signals applied in the experiment are illustrated in Fig. 11.1. The real waveform of the stimuli was affected by the phase of components, so that the in-phase and random-phase stimuli had different waveforms. The in-phase waveform had remarkable periodic peaks corresponding to the F_0. For the random-phase condition, each component was compounded with different phases so that the waveforms had no such significant periodic four peaks.

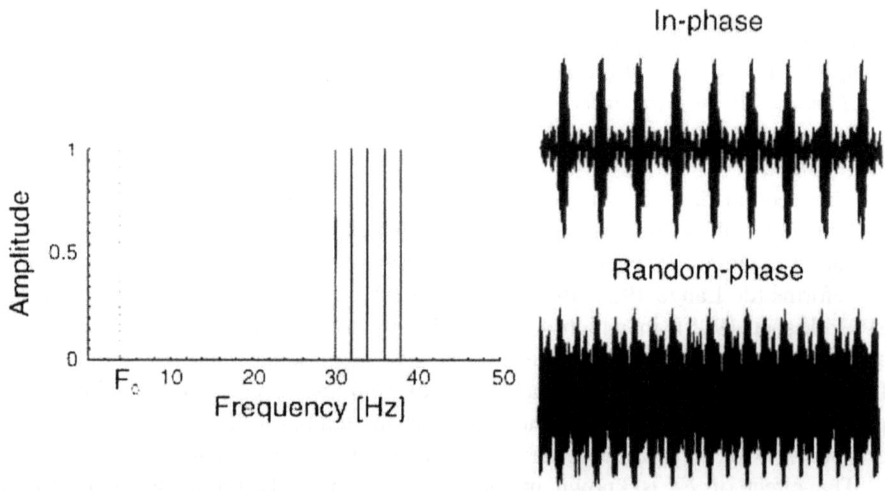

Fig. 11.1 An example of the spectrum of the complex signal used in the experiment. *Left* Complex components are 30, 32, 34, 36, and 38 Hz, where the energy of the fundamental frequency ($F_0 = 2$ Hz) is absent. *Right* Real waveforms in conditions of in-phase with remarkable peaks corresponding to the F_0 (*above*) and of random-phase without such clear periodic peaks of F_0 (*below*)

The subjective flicker rate of the stimulus was obtained by means of the "method of limits." The flicker with compound waveforms was used as the test stimulus, and sinusoidal flicker was used as a reference stimulus. These two stimuli were presented in pairs with a blank interval. The observers' task was to judge which of these two stimuli seemed to flicker at a faster rate. As the reference stimulus, we used ascending and descending series. That was, the comparison stimulus was varied in steps, from a low frequency to a high frequency (or vice versa) to measure the value at which the observers' response reversed. The mean of the two values before and after reversal of the observers' response was determined as the matched frequency of the test stimulus. When the observers perceived two or more rates for one test stimulus, they were asked to judge with the rate perceived most strongly. This means that the observers matched the sinusoid to the most prominent component of the compound waveforms; and thus, one matched frequency was obtained through one trial. Intervals of the comparison stimulus were 0.1 Hz step for frequencies below 1, 0.2 Hz step for 1–3 Hz, and a 1 Hz step for above 3 Hz. In the descending series, trials started from a value a few Hz above the highest frequency of the components in the test stimulus. There were two series of the comparison stimulus (ascending and descending) and two orders of presentation (test-comparison and comparison test), giving a total of four conditions. For each condition four trials were repeated. Thus, 16 matched frequencies were obtained for each test stimulus.

Results of the probability of responses to each matched frequency are shown in Fig. 11.2 as a histogram. For the in-phase stimuli, observers perceived the rates at F_0. This frequency is easily detected, because it is consistent with the time interval between the periodic peaks appearing in the temporal waveforms as shown in the right of Fig. 11.1. For the random-phase stimuli, matched frequencies were comparable to the several aperiodic peaks which correspond to the component frequencies. We could detect the flicker rates from local peaks in the waveforms in this low frequency range only (3–7 Hz). In the high frequency range, however, the fundamental frequency F_0 was perceived most frequently for both in- and random-phase stimuli, which is called the missing fundamental phenomenon, even allowing some exception such as certain multiples of F_0.

Figure 11.3 shows the observers' responses within $(1 \pm 0.1)F_0$ as a function of the fundamental frequency F_0. Both curves have a similar profile, except that the probability was about 10 % higher for the in-phase condition. Although probability was affected by the phase, the most frequently perceived rates were about F_0 in all cases. The highest probability is seen at $F_0 = 2$ and 2.5 Hz for the random and in-phase condition, respectively. These values correspond to the periods of 500 and 400 ms, which are similar to the "sensitive range," reported by Fraisse (1984). He reported that in the sensitive range (500–700 ms), the sensitivity increased to the periodicity of successive presentation of the stimuli. Our observers might also have responded sensitively to the periodicity of the flickering stimuli in this range. Thus, observers may detect the rates at fundamental frequency, which are not included in the power spectrum of the stimuli. One promising operation that gives the phase-independent prediction for our empirical evidence is the ACF (Fig. 11.4). Actually,

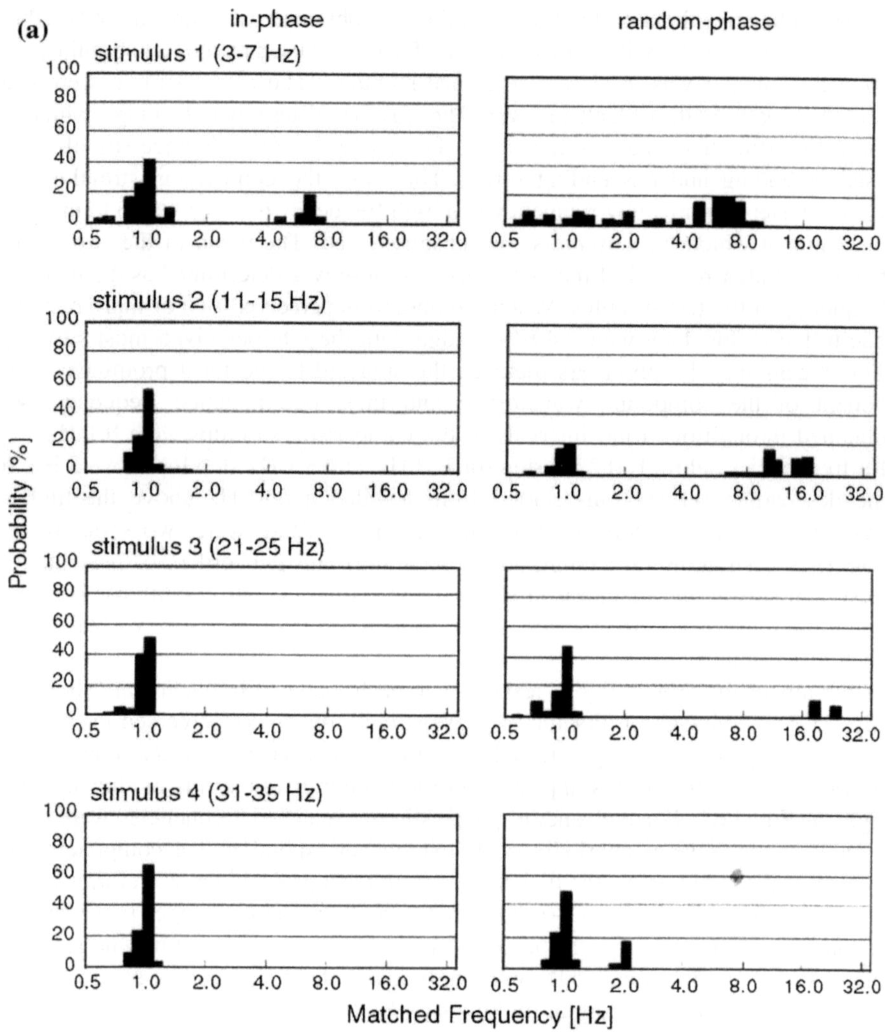

Fig. 11.2 Results of the response probability to matched frequencies, in-phase and random phase conditions with four subjects. **a** $F_0 = 1$ Hz with different frequency components. **b** $F_0 = 0.75$, 2, 2.5, and 3 Hz

the ACF of the real stimulus waveforms had identical profiles for both phase conditions used in the experiment. This result is consistent with the fact that the ACF has particular peaks corresponding to the F_0. It is possible to suppose a mechanism to extract the periodicity at F_0 from the peaks in the ACF of the stimulus. In the experiment, the observers' response at F_0 was slightly affected (about 10 %) by phase (Fig. 11.3), and some responses were seen at multiples of F_0 with the random-phase stimuli (Fig. 11.5).

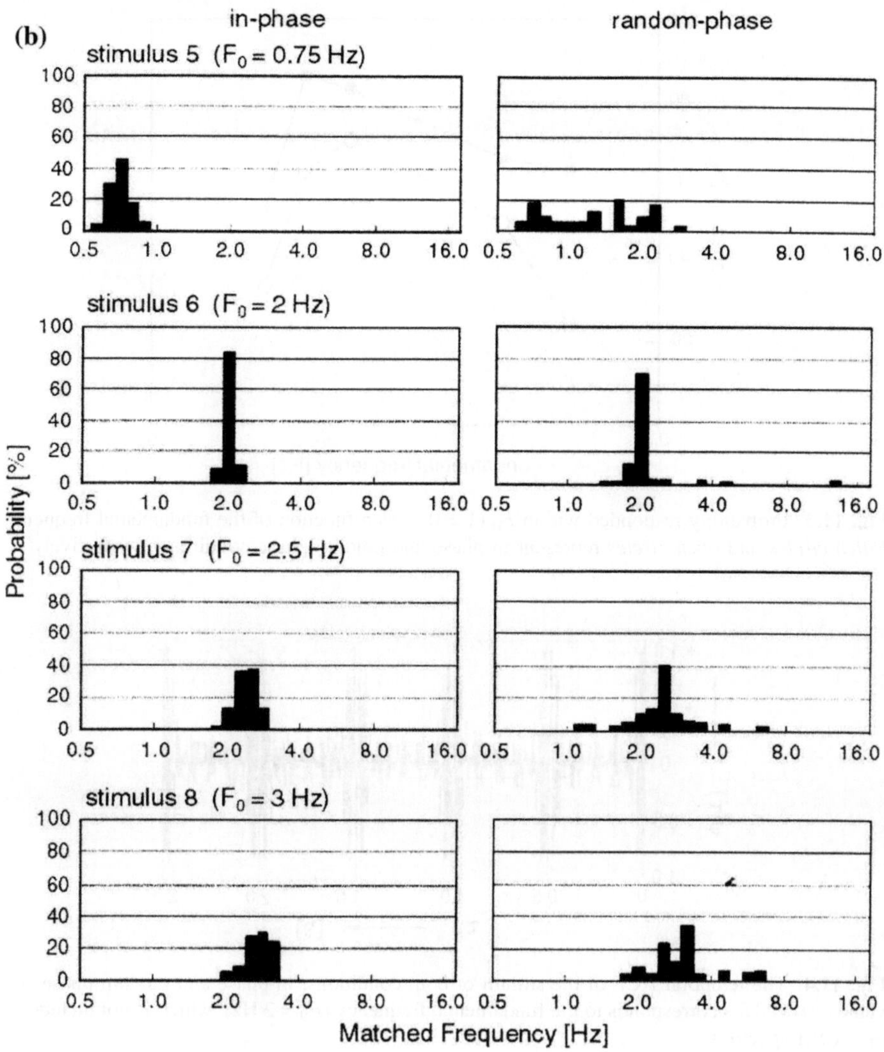

Fig. 11.2 (continued)

11.2 Preferred Conditions of a Flickering Light

If the flicker light like a twinkling stars with a certain degree of fluctuation may be utilized on the stage of opera house, we are interested in knowing a more preferred condition than perfectly periodic sinusoidal one.

In order to obtain basic design data for the temporal design, subjective preference judgments of the flickering light have been conducted (Soeta et al. 2002a).

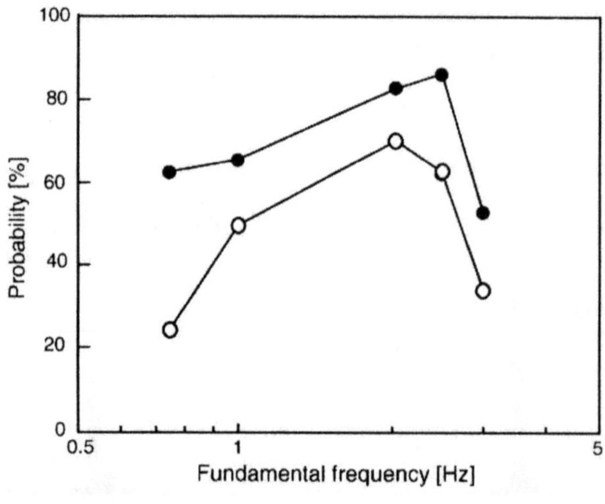

Fig. 11.3 Probability responded within F_0 (1 ± 0.1) as a function of the fundamental frequency. *Filled circles* and *open circles* represent in-phase and random-phase conditions, respectively

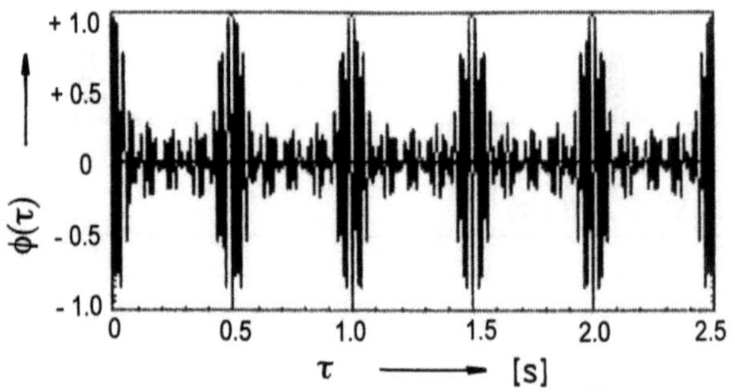

Fig. 11.4 The temporal ACF of the stimuli of both conditions, in-phase and random-phase. The value of $\tau_1 = 0.5$ s corresponds to the fundamental frequency ($F_0 = 2$ Hz), which is not included in the power spectrum

First of all, it has been found that the preferred sinusoidal period of the flickering light was about 1.0 s.

In order to attain a more preferred condition, a certain degree of fluctuation in the flickering light was introduced such as the twinkling stars, in addition to the preferred flickering sinusoidal period of 1.0 s. In this procedure, the factor ϕ_1 shown in Fig. 11.6 can be controlled, which is extracted from the ACF of the time varying signal of flickering light.

The most preferred fluctuations of the flickering light for individual subjects and global ones have been obtained as described by the factor ϕ_1 as indicated in Table 11.1. The value of ϕ_1 is known as a "pitch strength" in the sound signals

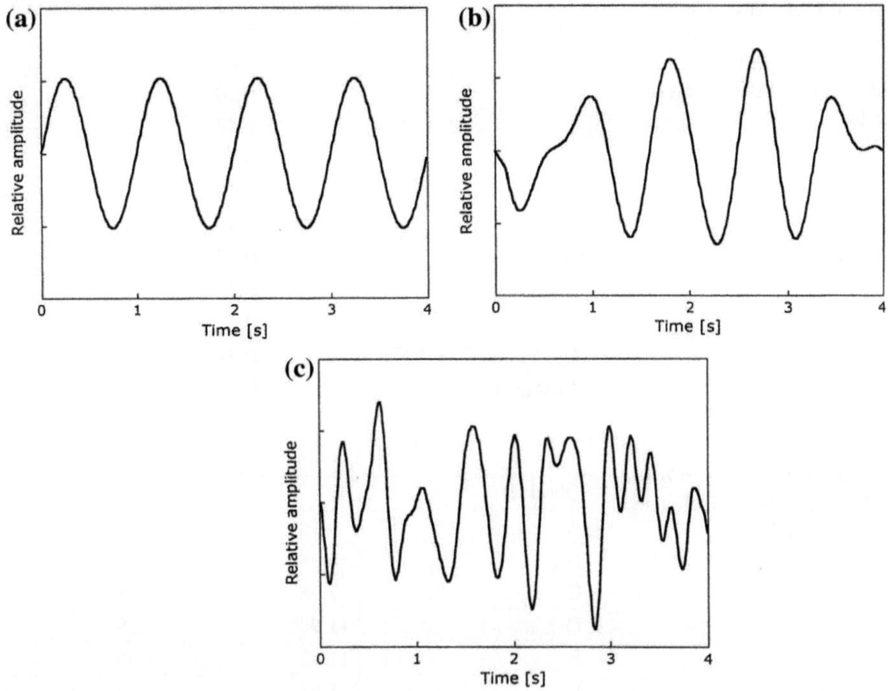

Fig. 11.5 ACF examples of stimuli used. **a** Sinusoidal wave. **b** Δf = 1 Hz. **c** Δf = 4 Hz

Fig. 11.6 The temporal ACF of a stimulus, and the definitions of ϕ_1 and τ_1

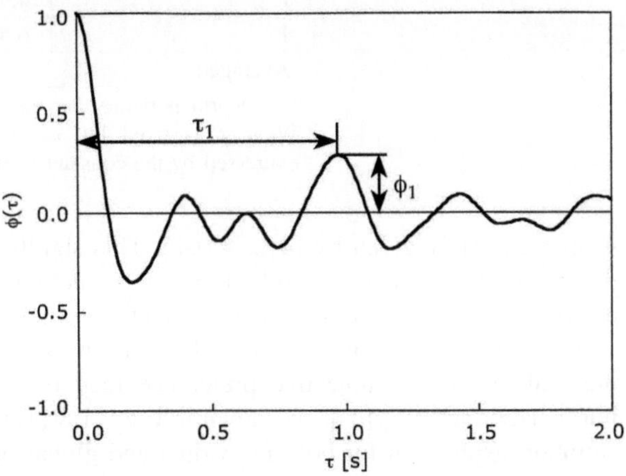

(Ando 2009a). In music performance, it is regarded as a kind of artistic expression in the temporal domain.

The most preferred conditions of individuals are distributed in a range of $[\phi_1]_p = 0.27{-}0.90$ (Soeta et al. 2005). The averaged value for a number of subjects

Table 11.1 The most preferred condition of the flickering light $[\phi_1]_p$ for each observer and the averaged value

Observer	$[\Phi_1]_p$
A	0.51
B	0.50
C	0.47
D	0.58
E	0.45
F	0.90
G	0.27
H	0.33
I	0.33
J	0.31
Averaged	0.46

Table 11.2 Values of α and β for each observer obtained for Eq. (11.1)

Observer	α	β
A	9.19	1.17
B	4.98	0.72
C	7.98	1.28
D	11.98	1.39
E	11.42	1.36
F	5.53	1.36
G	6.94	1.46
H	24.72	2.37
I	14.93	2.41
J	6.57	1.23
Averaged		1.47

It is worth noticing that the average value of β = 1.47 ≈ 3/2. When β is fixed by 3/2, the individual differences may be expressed by the constant α, (the averaged value of α is 10.98)

is approximately given by $[\phi_1]_p \approx 0.46$. This signifies that the extreme conditions $\phi_1 = 0$ (perfectly random) and $[\phi_1]_p = 1.0$ (perfectly periodic like the sinusoidal) are not preferred, but a certain degree of fluctuation is much more preferred.

After obtaining the most preferred condition for individual observers (Fig. 11.7), the scale value of subjective preference may be obtained as a function of the normalized factor $\phi_1/[\phi_1]_p$ as shown in Fig. 11.8 (Table 11.2). Thus, the following common expression for both individual and global observers yields

$$S \approx -\alpha|x|^{3/2}, \tag{11.1}$$

where $x = \log \phi_1 - \log\,[\phi_1]_p$. The weighting coefficient averaged for global subjects has been obtained, $\alpha \approx 11.0$. The behavior of scale value of subjective preference is similar to those of Eq. (6.10) for the sound field.

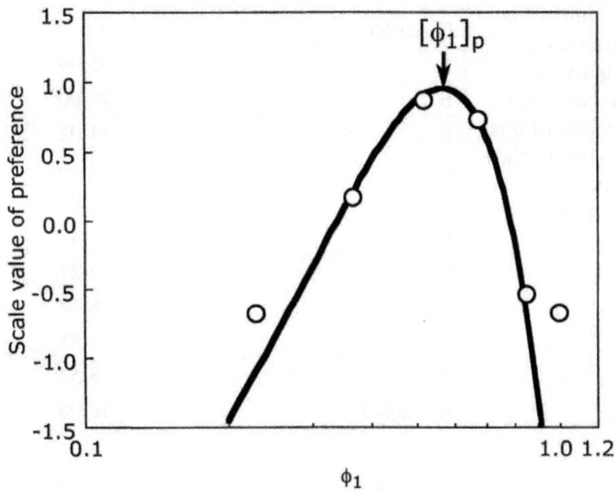

Fig. 11.7 An example of obtaining the most preferred, $[\phi_1]_p$ (≈0.58), for a single subject. The scale value at $\phi_1 = 1$ is not plotted in the curve fitted, because the decline of preference saturated already at $\phi_1 = 0.85$ in this case

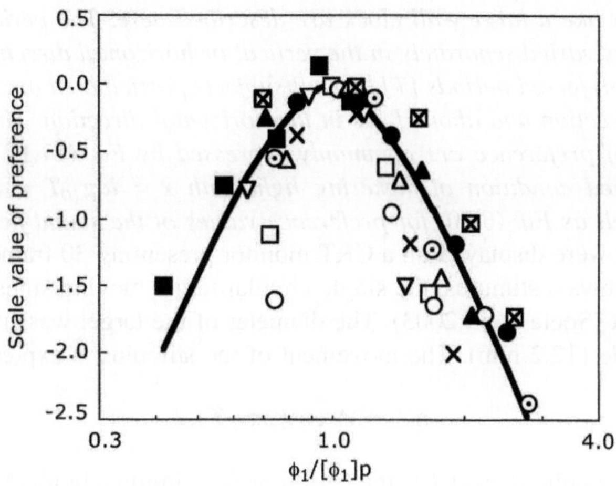

Fig. 11.8 Normalized scale values of preference for all subjects. The *solid curve* is calculated value by Eq. (11.1) with constants $\alpha = 10.98$ and $\beta = 3/2$ (Table 11.2). *Different symbols* indicate scale values obtained by different subjects. The abscissa ϕ_1 is normalized by $[\phi_1]_p$. The scale value at $[\phi_1]_p$ is adjusted to zero, without loss of any generality

It is assumed that in the thermal environment control, "breeze" could be introduced by application of the factor ϕ_1 which is extracted from the ACF of wind speed. Such breeze could be utilized in an opera house as well as daily environment in the summer time without consuming the large energy for the temperature control (Table 11.3).

Table 11.3 The value of α obtained for each observer representing the individual difference and the averaged value, when the value of β is fixed at 3/2 in Eq. (11.1), then individual differences may be expressed by the constant α

Observer	α
A	15.99
B	18.00
C	11.02
D	14.45
E	13.82
F	6.49
G	7.27
H	8.32
I	5.88
J	8.53
Averaged	10.98

11.3 Preferred Condition of Oscillatory Movements of a Circular Target

Preference judgments by applying the PCT for sinusoidal movements of a single circular target like a large wall clock are described here. The period of stimulus movements was varied separately in the vertical or horizontal direction. It is found that the most preferred periods $[T]_p$ for all subjects participated are about 1.0 s in the vertical direction and about 1.3 s in the horizontal direction. The curve of the scale values of preference are commonly expressed by Eq. (11.1) *similar to the above-mentioned condition of flickering light with $x = \log_{10}T - \log_{10}[T]_p$ and $b = 3/2$, as well as Eq.* (6.10) *for preference values of the sound field.*

The stimuli were displayed on a CRT monitor presenting 30 frames per second. Figure 11.9 shows a stimulus, the single circular target moving sinusoidal, used in the experiment (Soeta et al. 2003). The diameter of the target was subtended 1° of the visual angle (12.2 mm). The movement of the stimulus is expressed as:

$$\mathrm{h(t)} = A\,\cos(2\pi t/T), \tag{11.2}$$

where A is the amplitude and T is the period of the stimulus. In all experiments, the amplitude A was fixed at 0.61 cm on the monitor screen, corresponding to 0.5° of visual angle. The white target and black background corresponded to gray levels 40 and 0.5 cd/m^2, respectively. The monitor presenting the stimuli was placed in a dark room 0.7 m away from the subject's eye position to maintain natural binocular.

Subjective preference tests for the period of movements in the horizontal and vertical directions were examined separately. The period of stimulus movement T in Eq. (11.2) was varied at six levels: $T = 0.6$, 0.8, 1.2, 1.6, 2.0, and 2.4 s. Thirty pairs combining six different periods constituted each series, and ten series were conducted for all ten subjects in the experiments by the PCT.

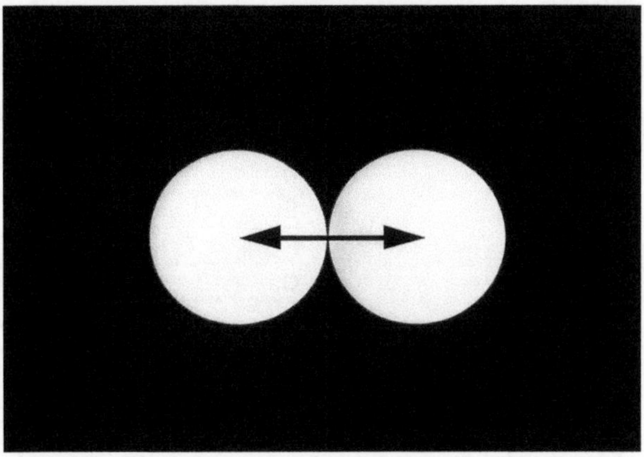

Fig. 11.9 Target of the horizontal movement used in the experiment

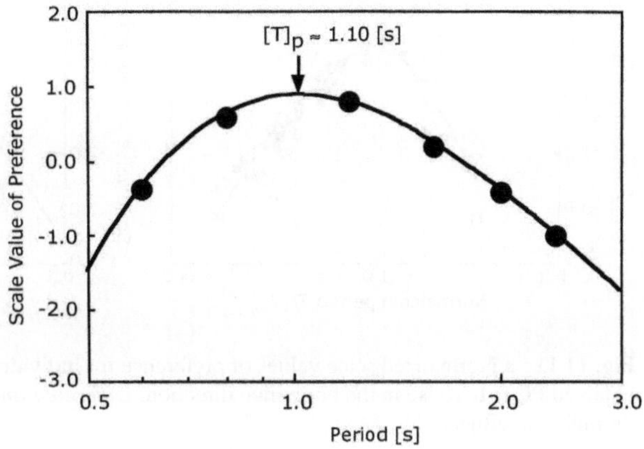

Fig. 11.10 An example of obtaining the most preferred period $[T]_p$ ($\approx$1.10 [s]), in the *vertical* direction (subject J)

The most preferred period $[T]_p$ for each subject was estimated by fitting a suitable polynomial curve to a graph on which scale values have been shown in Fig. 11.10. The peak of these curve denote the individual subject the most preferred values that are listed in Table 11.4. The individual values of preference for vertical and horizontal direction are plotted in Fig. 11.11a, b, respectively, so that preference curves may be expressed by Eq. (11.1), also.

The global value of the most preferred period was about 0.97 s for vertical movement, and about 1.26 s for horizontal movement as indicated by the averaged values in Table 11.4. All subjects indicated, therefore, preferred periods in the vertical direction were slightly shorter than that of those in the horizontal direction ($p < 0.01$).

Table 11.4 The most preferred periods $[T]_p$ of vertical and horizontal movements of the target for each subject and the averaged values

Subject	Vertical (s)	Horizontal (s)
A	1.15	1.28
B	1.05	1.82
C	0.78	1.31
D	1.16	1.79
E	0.85	0.91
F	0.83	1.05
G	1.08	1.31
H	0.81	1.04
I	0.93	0.98
J	1.10	1.13
Averaged	0.97	1.26

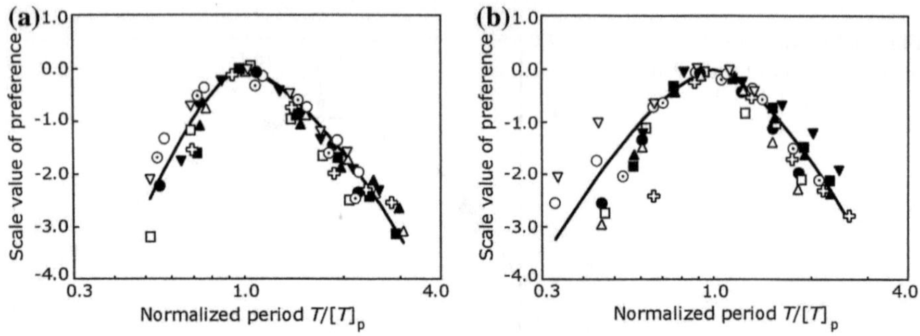

Fig. 11.11 **a** Normalized scale values of preference for individual subjects in the vertical direction obtained PCT. **b** Those in the horizontal direction. *Different symbols* indicate scale values obtained by different subjects

11.4 Matching Movement of Camphor Leaves with Acoustic Tempo

*In order to attain basic knowledge, the matching tests between images of camphor leaves on outer stage (ex. Fig. 12.6) with the periodical of the sound pulse was performed. Subjects watched the displayed image of camphor leaves moving in different speeds of wind while simultaneously listening to the periodical of the sound pulse changing the period ranging from 0.08 to 1.28 s. Results show that the matching period of the sound pulse is about half of the delay time at the first peak of the*ACF (τ_1) *which is analyzed by the gray level of moving-leave images. Note that the speed of the wind is not the temporal factor to describe the sensation for the camphor leaves movement.*

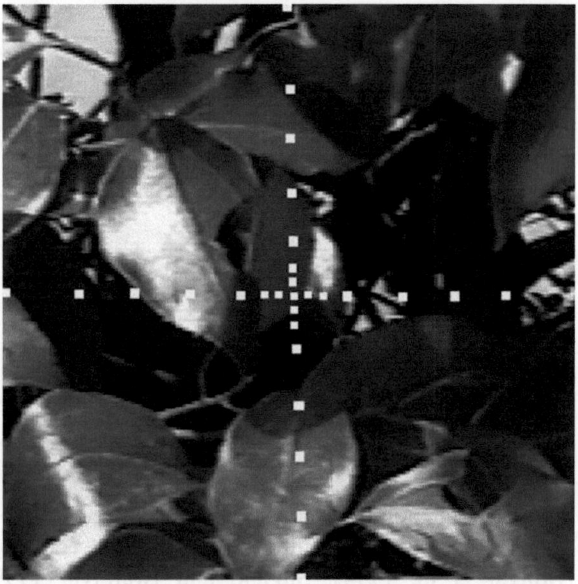

Fig. 11.12 Camphor leaves used in this study; white dots show 29 positions analyzed

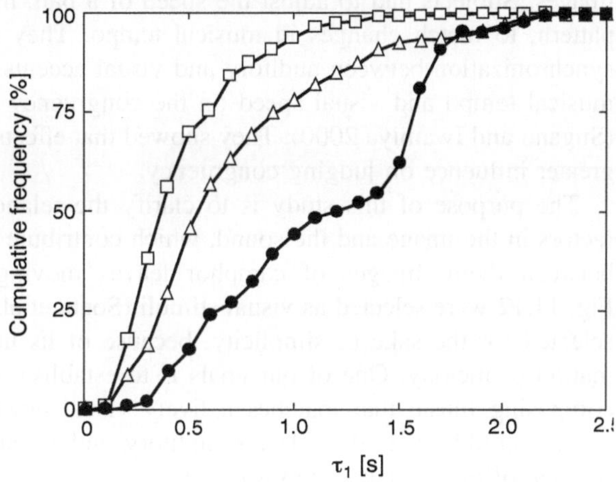

Fig. 11.13 Cumulative frequencies of values of τ_1 obtained at 29 positions shown in Fig. 11.12. Average wind speeds represented by: *filled circle* 0.7 s, *open triangle* 2.4, and *open square* 3.1 [m/s]

The visual and auditory systems provide us with the majority of the information that we are receiving in an opera house. A number of studies have dealt with the relationship between audio and visual perception. For example, Gebhard and Mowbray (1959) and Myers, Cotton and Hilp (1981) investigated matching repeating tone bursts to light pulses. To extend the knowledge regarding the interactions between sound on the impression of video and of video on the impression of sound, some experiments used audiovisual media (Bolivar et al. 1994; Lipscomb and Kendall 1994; Iwamiya 1994). These studies are mostly concerned with the degree of matching and its evaluative dimension. The ability to

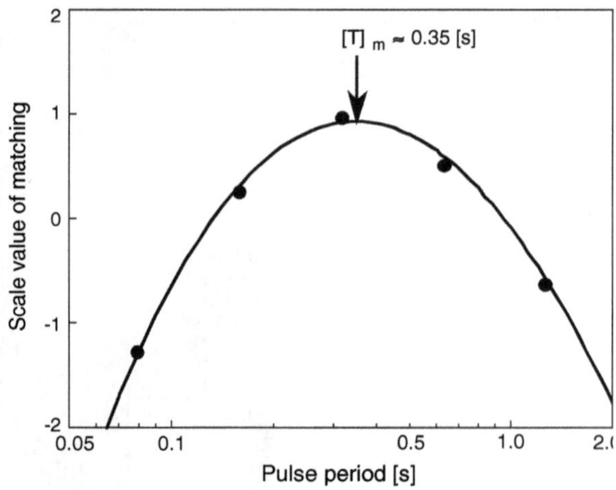

Fig. 11.14 Example of obtaining most-matched-pulse period $[T]_{\mathrm{m}} = 0.35$ s with movement of camphor leaves (Subject E, wind speed ≈ 2.40 m/s)

interpret an action film depends on the combination of semantic (i.e., meaning) and formal (e.g., temporal) information flowing across auditory and visual channels (Bolivar et al. 1994). Sugano and Iwamiya (1999) investigated the temporal information in visual motion to determine the congruency of music and moving images. Subjects had to adjust the speed of a ball, moving in a circular or square pattern, to match changes in musical tempo. They also examined effects of the synchronization between auditory and visual accents and those matching between musical tempo and visual speed on the congruency of motion picture and music (Sugano and Iwamiya 2000). They showed that effects of the synchronization had a greater influence on judging congruency.

The purpose of this study is to clarify the relationship between the temporal factors in the image and the sound, which contribute to perceptions of congruency between them. Images of camphor leaves moving in the wind as shown in Fig. 11.12 were selected as visual stimuli (Soeta et al. 2001). The sound pulse was selected for the sake of simplicity, because of its uncomplicated semantic information or melody. One of our goals is to establish a method for selecting and/or composing music that matches a lively visual environment on a stage. Such a method could be used to design auditory and visual environments in which the passage of time is taken into account.

We selected three five-second images at three different wind speeds, so the 0.71 ± (0.03), 2.40 ± (0.08), and 3.12 ± (0.29) m/s. The cumulative frequencies of the τ_1 at 29 positions are shown in Fig. 11.13. The median (50 %) value of τ_1 was 0.33 s at a wind speed of 3.12, 0.53 s at a wind speed of 2.40, and 1.12 s at a wind speed of 0.71 m/s.

Five-second sequential pulses, 0.08, 0.16, 0.32, 0.64, and 1.28 s with each pulse-width of 63 μs, were applied for the matching tests. Seven subjects, 21-to 24-years old and having normal hearing and binocular vision, participated in this study. The monitor presenting the visual stimuli was placed in the front of subjects' eye position at a distance of 1.1 m to keep foveal fixation (natural binocular). The loudspeaker was

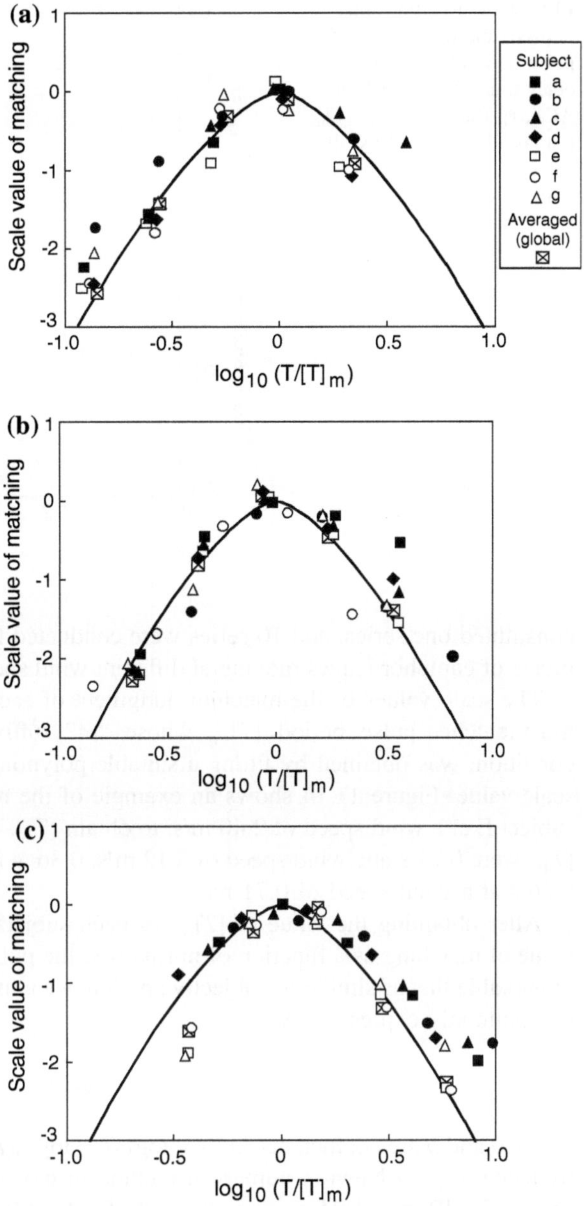

Fig. 11.15 Scale values of matching as a function of the period of pulse sounds normalized by most-matched-pulse period $[T]_m$ at **a** wind speed ≈ 0.71 m/s, **b** wind speed ≈ 2.40 m/s, and **c** wind speed ≈ 3.12 m/s

placed on the monitor. The subjects' head and eye positions were unconstrained. The sound pressure level at the center position of the subject's head was kept constant at a peak of 78 dBA. The subjects judged which sound pulses more subjectively matched up with the movement of camphor leaves. Ten pairs of combined 5-level periods

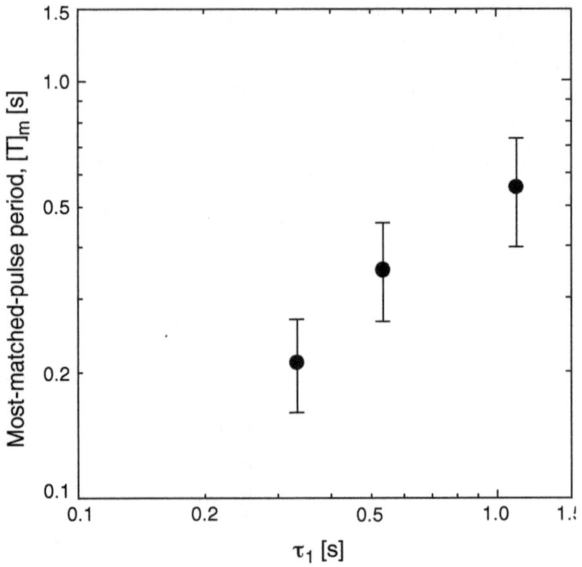

Fig. 11.16 Relationship between the most-matched-pulse period $[T]_m$ and the delay time of first peak of ACF, τ_1. Ranges of the $[T]_m$ graphically obtained by all subjects at 0.1 below the maximum matching score

constituted one series, and 10 series were conducted for all seven subjects for each image of camphor leaves moving at different wind speeds.

The scale values of the matching judgment of each subject were obtained. The most-matched-pulse period $[T]_m$, whose "m" suffix denotes the most-matched condition, was obtained by fitting a suitable polynomial curve to the graph of the scale value. Figure 11.14 shows an example of the matching evaluation curve for subject E at a wind speed of 2.40 m/s, to obtain $[T]_m \approx 0.35$ s. Global values of the $[T]_m$ were 0.21 s at a wind speed of 3.12 m/s, 0.36 at a wind speed of 2.40 m/s, and 0.56 s at a wind speed of 0.71 m/s.

After obtaining the value of $[T]_m$ for each subject, Fig. 11.15 shows the scale value of matching as a function of normalized the pulse period for all subjects. It is remarkable that as similar to subjective preference a matching evaluation curve may be commonly expressed by

$$S = S_L \approx -\alpha|x|^{\beta}, \quad (11.3)$$

where α and β are coefficients and $x = \log_{10}T - \log_{10}[T]_m$. The values of β estimated by using a quasi-Newton numerical method in global values were 1.18 at a wind speed of 3.12 m/s, 1.94 at a wind speed of 2.40 m/s, and 1.43 at a wind speed of 0.71 m/s. The average value of β was 1.52 (≈3/2) here, thus it may be fixed at 3/2 also. The solid line in Fig. 11.15 indicates the matching curves represented by Eq. (11.3) with $\beta = 3/2$. The characteristics of the matching curve can be approximately expressed by the single coefficient α, which represents the sharpness of the matching curve.

Fig. 11.17 Two-dimensional spatial textures applied with different values of ϕ_1 applied for the subjective preference judgment

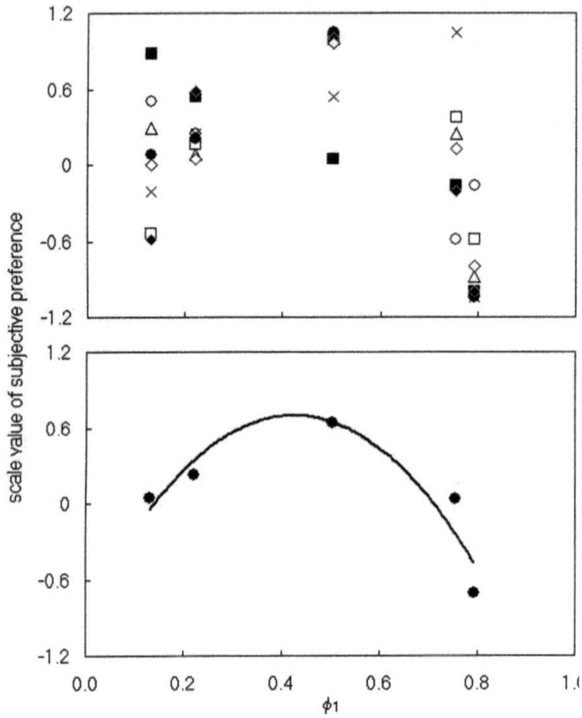

Fig. 11.18 **a** Scale values of preference for individual subjects. **b** Averaged values. For averaged values, a fitting curve is obtained with the 3/2 power of ϕ_1 in Eq. (11.1)

Figure 11.16 shows the $[T]_m$ as a function of the factor τ_1. A remarkable finding is that the matched period of sound pulses is about half of τ_1. The other factors, $\Phi(0)$ and ϕ_1, were not related to the most-matched-pulse periods in this experiment. It has been discovered that the delay time of the first peak of the ACF, τ_1, and is closely related to the perceived "pitch" of the lighting light (Sect. 11.1).

The weighting coefficient β in Eq. (11.1) is nearly equal to 3/2. This is consistent with the preference judgment for the sound field and the flickering light (Sects. 11.1 and 11.2). Equation (11.1), which represents the matching evaluation curve in the present study, and corresponds to the preference evaluation curve of the sound field, which could mean that the theory on the subjective preference of sound field might also be applicable to studies such as the congruency of music and motion sequences.

So far, results show that the matching period of sound pulses is roughly half of the delay time of first peak of the ACF, which is analyzed by the gray level of image.

11.5 Subjective Preference of Texture

To evaluate subjective preference of texture, an experiment was conducted by the paired-comparison test (PCT). By averaging the scale values of all subjects, it is found that the most preferred value of ϕ_1 for texture regularity extracted from the

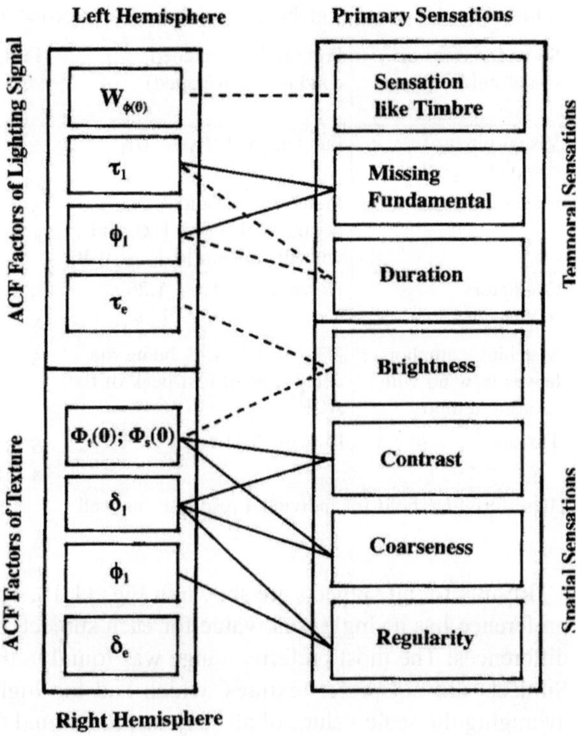

Fig. 11.19 Sumarization of temporal and spatial visual sensations (percepts) in relation to the factors extracted from the temporal ACF and the spatial ACF, respectively

spatial ACF is approximately given by $[\phi_1]_p \sim 0.41$. According to this result, a suggestion is made here that artistic expressions of color modulation and sequential form in drawings could be applied for the stage design.

The spatial factors are extracted from the two-dimensional ACF (Fujii and Ando, unpublished). To compare the degree of periodicity, the amplitude of the first peak in the ACF, ϕ_1 was considered under the condition of a roughly constant visual pitch δ_1 (Note in the case of sound signals this factor has been defined by τ_1 also, see Fig. 11.6). If the spacing between the objects in the pattern is equal, the calculated ACF does not decay that means a texture is, theoretically, perfectly regular. However, if the materials contain a kind of fluctuation, of the object size and spacing, and non-uniformity of the light reflection then the ACF gradually decay, and the value of ϕ_1, which is a measure of perceived regularity, could be considered as the measure for the degree of fluctuation in texture.

Ten, 22-to 24-years old subjects participated in the experiment. All had normal or corrected-to-normal visual acuity. Stimuli were presented on a display under a dark surrounding. The display was set at a distance of 1.5 m from the subjects. Subjects were presented pairs of two stimuli and asked to judge, which they preferred (PCT). All possible pairs from the five selected stimuli as shown in Fig. 11.17 were presented in a random order in one session. All subjects conducted ten series of sessions, giving a total of 100 judgments.

Table 11.5 Summary of global preferred or matched conditions of visions and visual fields

Sort of visions or visual fields	Preferred or matched conditions (averaged)	Formulae expressing the scale values	Section number described
Visual pitch of complex signal	Fundamental, $F_0 \sim 1/\tau_1$	–	11.1
Flickering light	Preferred sinusoidal period ~ 1.0 s (1 Hz), and with fluctuation $[\phi_1]_p \sim 0.46$	$S \approx -\alpha\|x\|^{3/2}$ where $x = \log_{10} \phi_1 - \log_{10} [\phi_1]_p$	11.2
Oscillatory movement	Horizontal $[T]_p \sim 1.26$ s Vertical $[T]_p \sim 0.97$ s	$S \approx -\alpha\|x\|^{3/2}$ where $x = \log_{10} T - \log_{10} [T]_p$	11.3
Matching camphor leaves in wind with acoustic tempo	$[T]_m \sim 1/2\ \tau_1$, τ_1 being the delay time of first peak of the ACF	$S \approx -\alpha\|x\|^{3/2}$ where $x = \log_{10} \underline{T} - \log_{10} [T]_m$	11.4
Texture	$[\phi_1]_p \sim 0.41$	$S \approx -\alpha\|x\|^{3/2}$ where $x = \log_{10} \phi_1 - \log_{10} [\phi_1]_p$	11.5

These formulae hold for individual responses as well

Results for all subjects are shown in Fig. 11.18. The scale value of the subjective preference has a single peak value for each subject, even allowing some individual differences. The most preferred range was found in the value of ϕ_1 for each subject. Subjects did not prefer textures, which had too high or too low values of ϕ_1. By averaging the scale values of all subjects, it is found that $[\phi_1]_p \sim 0.41$ was the most preferred value for texture regularity. The coefficients in Eq. (11.1) with the number of subjects was $\alpha \approx 3.9$, and $\beta = 3/2$.

It is worth noticing that the most preferred subjective preference of the flickering light with fluctuation as described in Sect. 11.2 is similar, $[\phi_1]_p \sim 0.46$ as well. It is considered, therefore, that a certain degree of fluctuation in both temporal and spatial factors is a visual property affecting subjective preference.

Preferred and matched conditions of visions and visual fields are summarized in Table 11.5. In concluding this chapter for further applications of vision, primary temporal and primary spatial sensations (percepts) are shown in Fig. 11.19.

Chapter 12
Design Theory of Opera House Stage Persisting Individual Creations

If the first stage of human life is the life of the body, and the second stage the life of the mind, then the third stage is the life of personality or ideas and creations that persist in social memory long after their individual creators have passed on. The parents by blood are father and mother for both the first and second stages for the gestation of about 40 weeks, however, the unique third stage is founded by the nature from the beginning of the universe. Thus ideas created by individual personalities, communicated to others, and then enter human culture live for a long time.

A general strategy for design is to characterize what humans experience (percepts) and what they prefer to experience, and to optimize their environments so as to realize their preferences. Percepts involve time and space, such that temporal and spatial factors determine different sets of experienced qualities. Temporal design of opera house optimizes temporal factors, spatial design, spatial factors. Temporal factors appear to be processed predominantly in the left cerebral hemisphere, whereas spatial factors are lateralized on the right. Design for creativity is considered in terms of temporal and spatial factors (Ando 2009a, b, 2013). The temporal criteria drive the left hemisphere; for example, acoustic parameters related to voice, speech, and music as well as visual temporal patterns related to movements such as leaves in a gentle breeze or twinkling stars. Factors related to spatial patterns, such as those generated in painting and sculpture, engage the right hemisphere.

Creativity in science and art arises from individual preference and sensibility. Such creative activities may keep a body in good health and a mind strong, no matter what the age, even up to the last moment of life.

12.1 Design Theory of Opera House Stage

It has been said that a healthy body relates to a healthy mind. We have therefore typically believed that there are only two stages of human life, that of the body (the first stage of human life) and the mind (its second stage). It is obvious that these two stages are also common to animals. If all of people believe only these two stages, no

Y. Ando, *Opera House Acoustics Based on Subjective Preference Theory*,
Mathematics for Industry 12, DOI 10.1007/978-4-431-55423-3_12

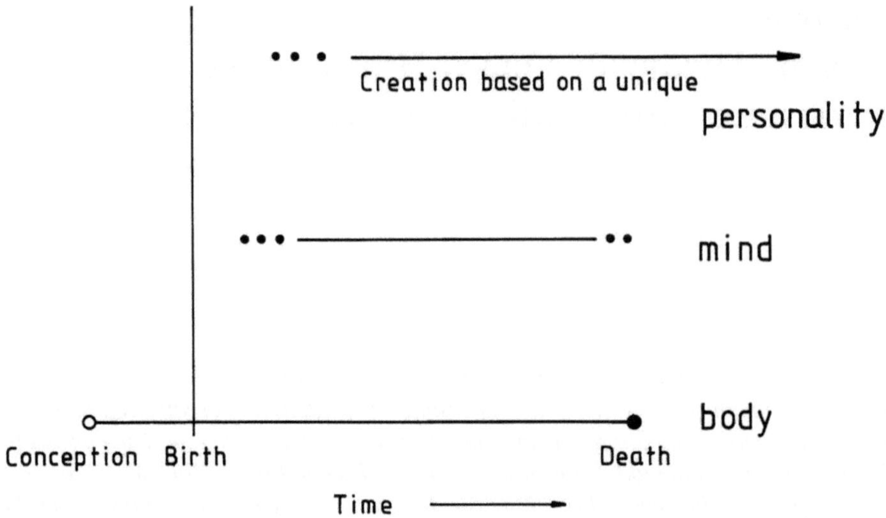

Fig. 12.1 Three stages of individual human life; the first (body), the second (mind), and the third (creations from personality are pledge of affection to this world from individuals)

cultural activities could be developed and discord each other always occur. However, there is also a third stage of life in which the creations of individual human beings persist after their first two stages have passed (Fig. 12.1). In this way, the works created may live on even after the end of the biological and mental life of their individual creators. Persons hope to leave something good behind them. Money is often left to others, but can lead to legal disputes amongst beneficiaries. On the other hand, unique, non-monetary individual creations can become integrated into ongoing, evolving common human cultures, thereby benefitting human society as a whole.

In order to realize development of a third stage of life, individual creations must be nurtured. Each of us begins life with a genetic endowment, a set of DNA, which can be considered as "a kind of seed," as shown in Fig. 12.2. It has been often said that the "soul" or "psyche" of a child of 3 years old, persists throughout life, to even a 100 years old. Thus, after birth it is very important to nurture each individual seed by designing its environments appropriately, by optimizing the various spatial and temporal factors in accordance with individual purpose of life and preference. This is the process that best maintains life.

The same is true for plant life—if the environment is well designed according to preference, such that the relevant, essential temporal and spatial factors are taken into account, then plant life thrives, and we can enjoy many of its products, such as the wonderful flowers in Fig. 12.3.

We have previously proposed a general theory of environmental design that uses observed human preferences to optimize physical factors that are psychophysically related to human perceptual qualities (Ando 2009a, b). Temporal factors involve physical parameters that determine stimulus qualities, such as the auditory qualities of

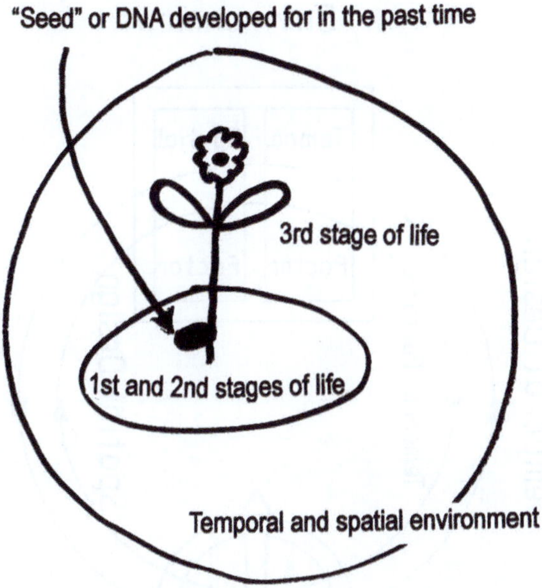

Fig. 12.2 Development of the third stage of human life originated from a genetic "seed" that is nurtured in an appropriate temporal and spatial environment. Due to preferred environment, personality for unique creation is well developed and effloresced like a flower of plant (Fig. 12.3)

Fig. 12.3 Lilies (called Casablanca) with over 90 flowers grown in their preferred environment (2006)

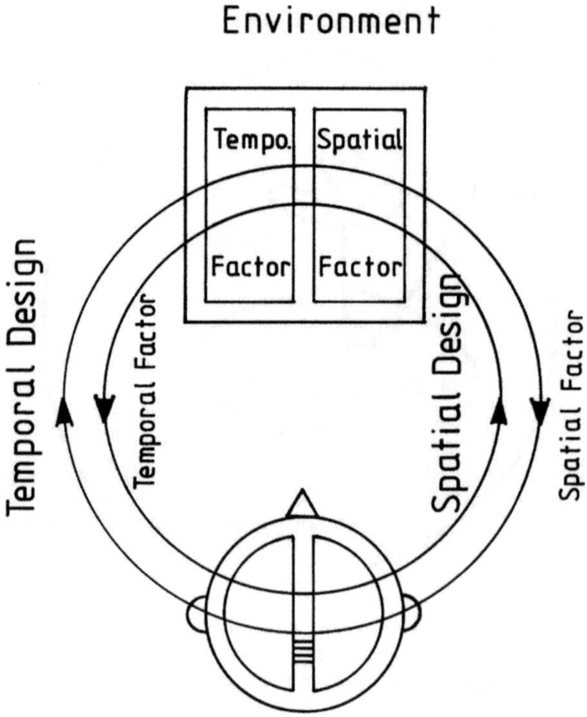

Fig. 12.4 Method of environmental design due to subjective preference theory (Ando 2009b). Temporal design involves optimization of temporal environmental factors that are typically processed in the left cerebral hemisphere, whereas spatial design is concerned with optimizing spatial factors that are processed in right cerebral hemisphere (Ando 2004, 2009a)

pitch, timbre, and loudness, and the visual qualities of texture, whereas spatial factors involve physical parameters that determine aspects of the stimuli associated with perceived locations in external space. We have investigated the observable neural correlates of perceptual factors and preferences, and have consistently found the temporal factors to be associated predominantly with the left hemisphere, and spatial factors associated with the right hemisphere, as shown in Fig. 12.4. We hypothesize that a preferred environment may activate both hemispheres, and this may optimally motivate creation. We believe that the satisfaction of subjective preferences in temporal and spatial realms always moves in the direction of maintaining life.

A well-designed environment and an individual's personality may resonate, and thereby play an important role for facilitating unique creative works that can then be shared with other people. A hypothesis pursued by a unique personality may expose an aspect of the world that had not yet been explored (Fig. 12.5). The set A^C in the diagram indicates an infinite number of unknowns to be solved by individuals. The hypothesis and the experience of the individual can be communicated to others through publication such that the knowledge can be shared socially and enter into culture (set A).

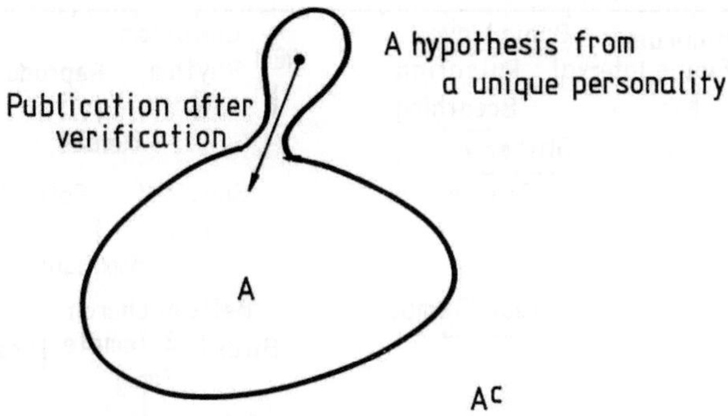

Fig. 12.5 Integration of individual explorations and creations into human social memory (culture) through publication after testing of verification. *A* Limited knowledge, which has been clarified and shared socially and then enter into culture. A^C Infinite number of unknowns. Unknown problems waiting to be solved and worked by individuals

12.2 Design Study of an Opera House

Opera is an epitome of the human life. People may enjoy performance on the stage being intertwined by the individual life receiving a message through drama, and the outside Nature as well. This includes affections from sum, moon, stars, water, wind, cloud, trees, and flower. These may act as flexibility of opera even if extraordinary weather change is occurred, individual creations may solve inducing in art and science. Since the time of big bang, the environment or universe have been dramatically developed with deep affection of Nature as resulted phenomena at present stage. Thus, the environment is life surrounded by full of affection of Nature that we are receiving it always. If we consider a longest period known as cyclic universe (Steinhardt 2009), environment in this world is still achieving a lot of miracle.

Stages of theatres together with natural activities has been known as Japanese NOH theatres and Javanese Gamelan theatres, as well as in ancient Greek and Rome theatres. In this section, a sketch of design study of four-seasonal opera house inducing individual creations is demonstrated here. Through participating opera, individuals might obtain an inspiration finding their personalities and thus creations accordingly.

12.2.1 Temporal Design

At the design stage of an opera house, first of all, the temporal design can be carefully considered (Ando 2009b, 2013; Ando and Criani 2014). The discrete periods in nature and human body as shown in Fig. 12.6, which are considered as the temporal design. The minimum period is order of 1 ms that corresponds to the

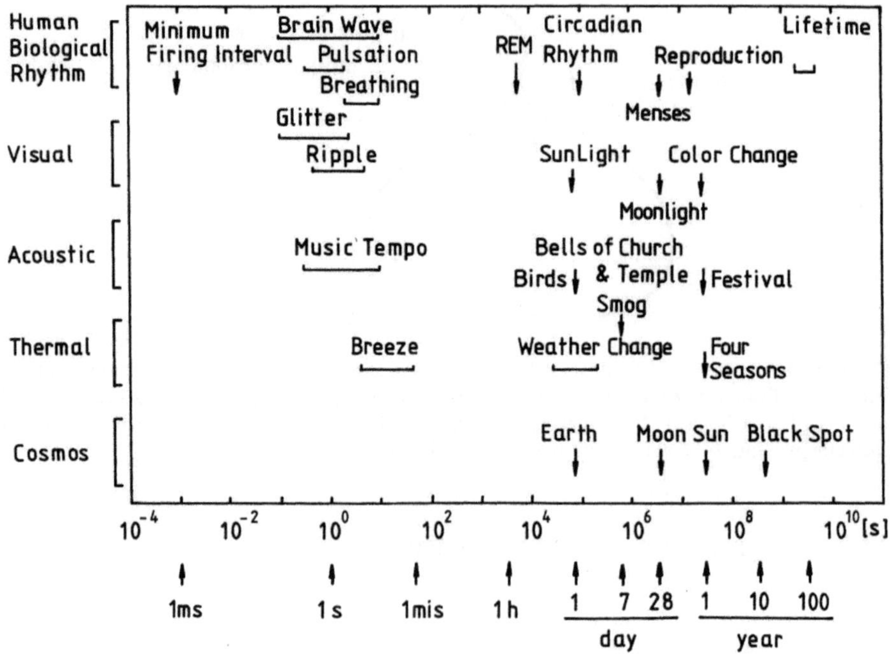

Fig. 12.6 Discrete and primacy existing periods. These periods may be considered in the temporal design of architecture and the environment. The periods are, for example, 1 ms, 1 s, 90 min (REM rhythm), 1 day, 1 week, 1 month, 1 year, 30 years (generation), 90 years (about life time)

neural firing rate, and the next distinct period is about 90 min. That relates to the rapid eye movement (REM) associated with brain rhythm of awaking and rest/sleep. This is an indication why we should have a short rest between programs of opera and drama. The period of 1 day is of most fundamental life.

Possibilities are the early morning opera beginning before sunrises, and sunset opera could be planned in addition to usual opera beginning in the evening, exactly at 7:00 pm. If every weekend, different programs of opera and drama performance are planed, then no additional costs for public notifications are required because of periodical performance that people know. Another possible program is like "moonlight opera" and "opera of four seasons" may be planed every full moon time in a month and every year, respectively.

It is worth noticing that human life has been usually for the first stage of body and the second stage of mind. However, for more joyful and peaceful life, it is to introduce the third stage of personality-oriented creations based on the individual "seed" (Fig. 12.2). If it could be included in opera performance for more varieties of human life stimulating the left and right human cerebral hemispheres for creating the temporal and the spatial environment, respectively (Chaps. 6 and 11). These three stages of human life, the temporal and spatial design may be considered in actual life as well as opera. Thus, all of these temporal design of opera house and

Fig. 12.7 **a**, **b** An example of cross-sectional sketches of opera house proposed hopefully to stimulating of individual creations. Walls are made of glasses blending the Nature. For example, possible opera utilizing three different stages, and finding the human life particularly individual potentiality for creations, as well as finding preferred environments for the third stage of human life toward peace avoiding bully that apart from the animal life with only the first and second stages of life centered on the concept of "time is money." *Note* worthy that "time is life."

human life may play important role for further possible creations of opera, drama, and others, so that more fruitful cultural life will be sprung out.

Furthermore, if many of people aware such an individual seed, then more idea will play avoiding being hard others and then wars. It has been impossible that these "social illness" based on only the first and second stages of life that is common to animals.

12.2.2 Crystal Opera House

As we can see the Delphi theater without any stage building and ceilings, so that drama was performed in such a open space. Since such ancient theaters, according to story in opera and drama being performing that the ceiling with upper roof and rear walls of audience floor could be opened and side walls are made of a strengthening double glazed. For example, audiences may look at shooting and twinkling stars that imagine a long story of universe as well. We might enjoy in certain situations with fresh air than entirely closed enclosure receiving deep affection from universe. Obtaining an idea enhancing survival even forthcoming big weather change, for example, we do hope solving such big problems by each personality of people living in this world. However, the concept of "time is money" or "economic animals" without the third stage of life may not realize sustainable environment and thus nor contributing toward a lasting peace.

Due to the temporal design and the preference theory of sound fields, a possible new type of crystal opera house may be proposed as shown in sketches of Fig. 12.7a, b. In order to stimulate creation in such an opera, and to obtain full variety of performance, three different stages, i.e., inner stage, upper stage, and outer stage of opera house together with Nature may be fully utilized.

Rear wall on the stage consists of large glass looking the outer stage and far scenery and overlooking the sea, for example, as indicated in sketches. Such a large glasses are utilized for side walls as well blending with the natural activities. A pit elevator is useful for adjusting level of the floor for orchestra performance and to enlarge the inner stage).

It is hoped that such a type of opera house may play important role for further creations according to individual personalities (DNA or seed), and integrate as a culture to be remained for a long time in this world and a lasting peace.

Appendix
Comparison Between Measured Orthogonal Factors Using a Dummy Head and Four Human-Real Heads

The purpose of measuring four orthogonal factors of the sound field is to examine calculated values done at the design stage or to examine sound quality based on the subjective preference theory. This kind of measurement is accumulated for improving design procedure and accuracy. In the acoustical measurement in an opera house, a human dummy head with two tiny microphones placed at two ear-entrances is often used as a receiver to obtain binaural impulse responses. It is interesting that real human heads are much convenient to carry from a seating position to others without excess baggage and cost.

When a human head is used, there may be individual differences in the measured factors due to the size and shape of the head, as well as the geometry of the body. Unconscious movement of the real head during the experiment may sometimes result in inaccurate data compared to the data obtained by a dummy head. However, in listening tests related to sound localization by individuals, judgment may be conducted quite accurately when individual human heads are applied in binaural recording (Sect. 3.3, Morimoto and Ando 1980; Nakajima et al. 1993). On the other hand, one of the advantages of using an artificial dummy head is the stability of the result, because the head can be fixed throughout the experiment.

To attain degree of differences in the acoustical parameters measured using human heads and a dummy head, measurements were conducted at typical locations in the stall of an opera house. The four orthogonal factors (LL, Δt_1, T_{sub}, IACC) and in addition factors (A, τ_{IACC}, and, W_{IACC}) were measured (Sakai et al. 2004).

In measurements in an opera house (Sect. 7.1), an omnidirectional dodecahedron loudspeaker was placed at the middle-front of the stage under the proscenium arch and 1 m away from the centerline. Four human real heads and one dummy head (Sennheiser 2002) were applied as receivers (ANSI 1985; Hidaka et al. 1995). The parameters are defined in Sect. 3.2. Dimensions of heads are defined in Fig. A1 and are listed in Table A1.

Y. Ando, *Opera House Acoustics Based on Subjective Preference Theory*, Mathematics for Industry 12, DOI 10.1007/978-4-431-55423-3

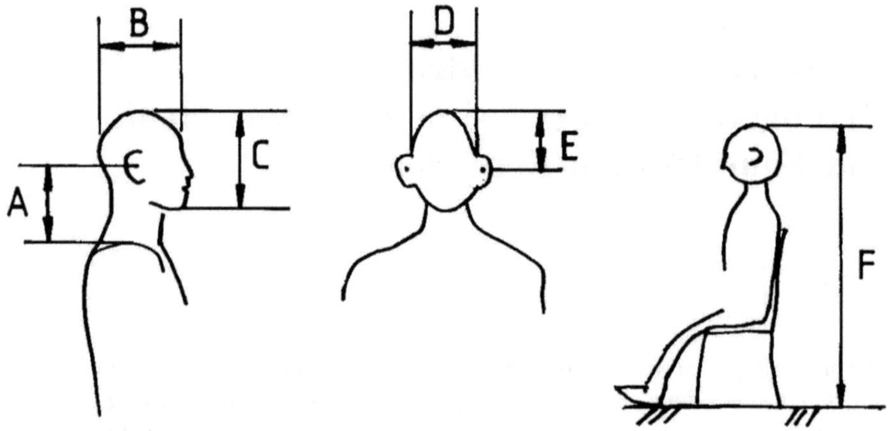

Fig. A1 Head-related dimensions measured, A through F (Table A1)

Table A1 Dimensions of the heads applied in measurement (cm)

Head	A	B	C	D	E	F
Dummy head	–[1]	18.0	22.0	15.0	12.0	132.0
Human head N	17.6	20.3	23.0	14.0	14.3	134.1
Human head A	19.0	21.0	21.0	14.5	15.0	129.0
Human head S	16.7	19.5	21.3	13.0	14.4	125.5
Human head H	14.7	20.1	22.6	13.3	14.0	124.5
Average for human heads	17.0	20.2	22.0	13.7	14.4	128.3

The dimension "F" includes the height of the seat (42.2 cm)
[1] No shoulder for Sennheiser head

The binaural impulse responses were measured applying the dummy head and the four human heads are shown in Fig. A2; quite a similar tendency was observed in the initial delay range (0–150 ms). The first reflection can be observed at about 50 ms. The measured relative LL, T_{sub}, and IACC are shown in Fig. A3. As can be seen in the error bars, for LL and IACC, results with different human heads were small enough, except for the frequency range of 4 kHz. The ranges between the maximum and minimum LL and IACC at 4 kHz were 4.1 dB and 0.16, respectively. The results of T_{sub} were close to each other measured with the four heads, as shown in Fig. A3b. The ranges measured of the factors are listed in Table A2.

So far, we found that the results for the dummy head agree closely with those for the four human heads, especially for terms of the temporal factors (Δt_1 and T_{sub}) where differences are almost negligible. However, some differences in the spatial factors (LL and IACC) were found in the frequency range centered on 4 kHz due to small difference in dimensions. From a practical point of view, these differences may be acceptable for the low frequency range below 2 kHz.

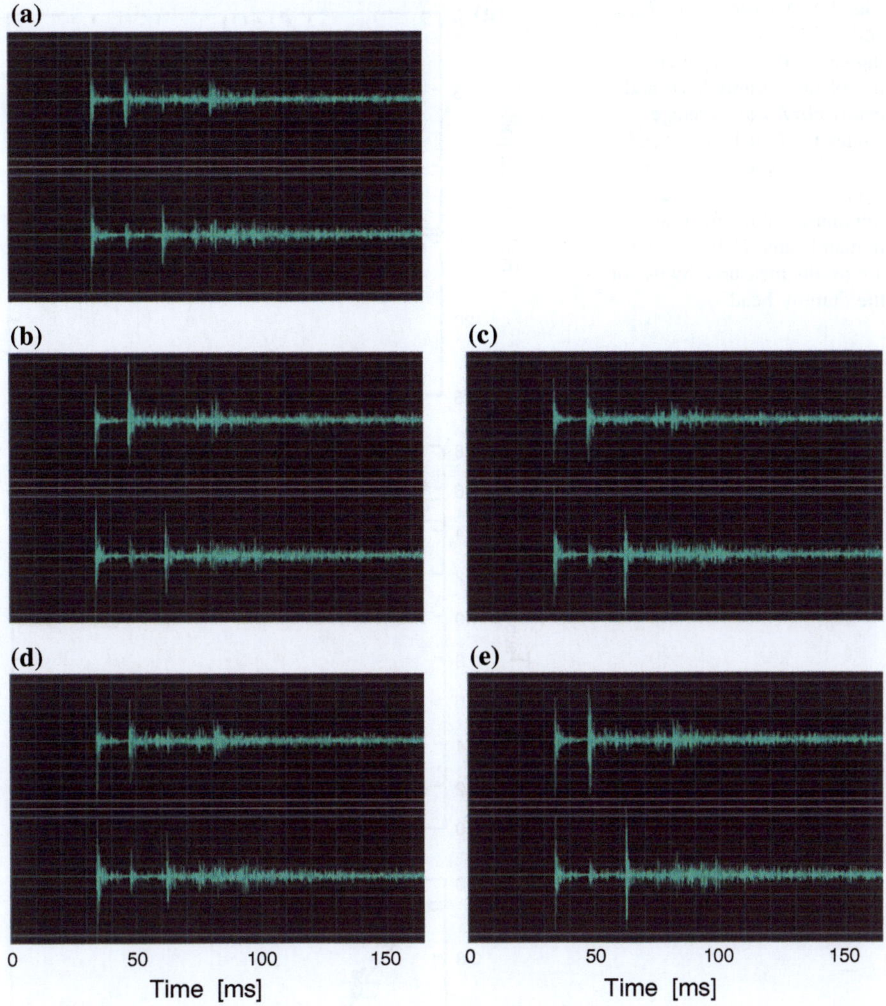

Fig. A2 Initial (0–150 ms) impulse responses measured by use of each head. **a** Dummy head. **b–e** Four human heads (N, A, S, and H). *Top: Left*-ear signal. *Bottom: Right*-ear signal

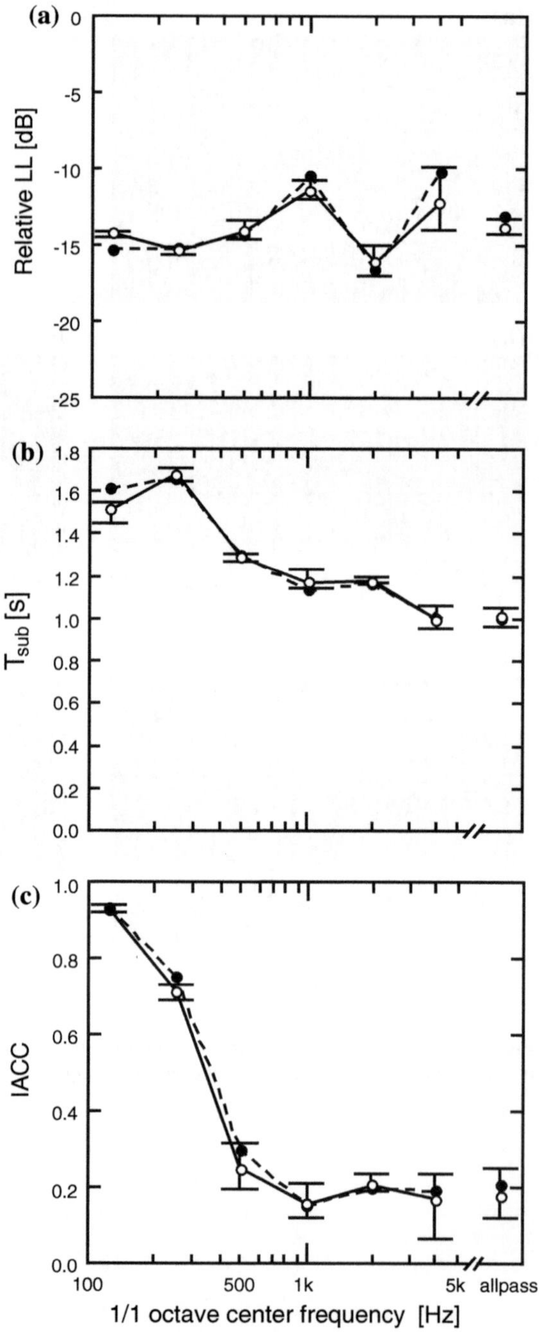

Fig. A3 Measured LL, T_{sub}, and IACC. *Full circles* indicate values measured by use of the dummy head and *empty circles* are average values for four human heads. **a** LL. **b** T_{sub}. **c** IACC. Error bars are the maximum and minimum values for four human heads. *Filled circles* are results measured by use of the dummy head

Table A2 The range (maximum–minimum) values of acoustic factors measured by the use of different human heads

Factor	Range
LL (500 Hz)	1.3 (dB)
LL (1 kHz)	1.4 (dB)
Δt_1 (left)	0.3 (ms)
Δt_1 (right)	0.4 (ms)
T_{sub} (500 Hz)	0.0 (s)
T_{sub} (1 kHz)	0.1 (s)
IACC (500 Hz)	0.11
IACC (1 kHz)	0.09
A value (left)	0.94
A value (right)	0.65
τ_{IACC}	0.09 (ms)
W_{IACC}	0.02 (ms)

References

Alrutz H (1981) Ein neuer Algorithms zur Auswerung von Messungen mit Pseudorauschsignalen, Fortschritte der Akustik, DAGA ’81, Berlin, pp 525–528

Ando Y, Shidara S, Maekawa Z, Kido K (1973) Some basic studies on the acoustic design of room by computer. J Acoust Soc Jpn 29:151–159 (in Japanese with English abstract)

Ando Y, Kageyama K (1977) Subjective preference of sound with a single early reflection. Acustica 37:111–117

Ando Y (1977) Subjective preference in relation to objective parameters of music sound fields with a single echo. J Acoust Soc Am 62:1436–1441

Ando Y, Imamura M (1979) Subjective preference tests for sound fields in concert halls simulated by the aid of a computer. J Sound Vib 65:229–239

Ando D, Gottlob D (1979) Effects of early multiple reflection on subjective preference judgments on music sound fields. J Acoust Soc Am 65:524–527

Ando Y, Morioka K (1981) Effects of the listening level and the magnitude of the interaural cross-correlation (IACC) on subjective preference judgment of sound field. J Acoust Soc Jpn 37:613–618 (in Japanese with English abstract)

Ando Y, Okura M, Yuasa K (1982) On the preferred reverberation time in auditoriums. Acustica 50:134–141

Ando Y, Alrutz H (1982) Perception of coloration in sound fields in relation to the autocorrelation function. J Acoust Soc Am 71:616–618

Ando Y (1983) Calculation of subjective preference at each seat in a concert hall. J Acoust Soc Am 74:873–887

Ando Y, Otera K, Hamana Y (1983) Experiments on the universality of the most preferred reverberation time for sound fields in auditoriums. J Acoust Soc Jpn 39:89–95 (in Japanese with English abstract)

Ando Y (1983) Calculation of subjective preference at each seat in a concert hall. J Acoust Soc Am 74:873–887

Ando Y, Hosaka I (1983) Hemispheric difference in evoked potentials to spatial sound field stimuli. J Acoust Soc Am 74(S1):S64–65(A)

Ando Y (1985) Concert hall acoustics. Springer, Heidelberg

Ando Y, Kurihara Y (1986) Nonlinear response in evaluating the subjective diffuseness of sound field. J Acoust Soc Am 80:833–836

Ando Y, Kang SH (1987) A study on the differential effects of sound stimuli on performing left- and right-hemispheric task. Acustica 64:110–116

Ando Y, Kang SH, Nagamatsu H (1987) On the auditory-evoked potentials in relation to the IACC of sound field. J Acoust Soc Jpn 8:183–190

Ando Y, Kang SH, Morita K (1987) On the relationship between auditory- evoked potential and subjective preference for sound field. J Acoust Soc Jpn 8:197–204

Ando Y (1988) Effects of daily noise on fetuses and cerebral hemisphere specialization in children. J Sound Vib 127:411–417

Y. Ando, *Opera House Acoustics Based on Subjective Preference Theory*,
Mathematics for Industry 12, DOI 10.1007/978-4-431-55423-3

Ando Y, Sakamoto M (1988) Superposition of geometries of surface for desired directional reflections in a concert hall. J Acoust Soc Am 84:1734–1740
Ando Y, Okano T, Takezoe Y (1989) The running autocorrelation function of different music signals relating to preferred temporal parameters of sound fields. J Acoust Soc Am 86:644–649
Ando Y, Yamamoto K, Nagamastu H, Kang SH (1991) Auditory brainstem response (ABR) in relation to the horizontal angle of sound incidence. Acoust Lett 15:57–64
Ando Y (1992) Evoked potentials relating to the subjective preference of sound fields. Acustica 76:292–296
Ando Y, Chen C (1996) On the analysis of the autocorrelation function of a-waves on the left and right cerebral hemispheres in relation to the delay time of single sound reflection. J Archit Plann Environ Eng Archit Inst Jpn 488:67–73
Ando Y, Johnson B, Bosworth T (1996) Theory of planning environments incorporating spatial and temporal values. Mem Grad Sch Sci Technol Kobe Univ 14-A:67–92
Ando Y, Singh PK (1996) A simple method of calculating individual subjective responses by paired-comparison tests. Mem Grad Sch Sci Technol Kobe Univ 14-A:57–66
Ando Y, Sato S, Nakajima T, Sakurai M (1997) Acoustic design of a concert hall applying the theory of subjective preference, and the acoustic measurement after construction. Acustica Acta Acustica 83:635–643
Ando Y, Noson D (eds) (1997) Music and concert hall acoustics, conference proceedings of MCHA 1995. Academic Press, London
Ando Y (1998) Architectural acoustics, blending sound sources, sound fields, and listeners. AIP Press/Springer, New York
Ando Y, Sato S, Sakai H (1999) Fundamental subjective attributes of sound fields based on the model of auditory-brain system. In: Sendra JJ (ed) Computational Acoustics in Architecture. WIT Press, Southampton
Ando Y, Sakai H, Sato S (2000) Formulae describing subjective attributes for sound fields based on a model of the auditory-brain system. J Sound Vib 232:101–127
Ando Y (2001) A theory of primary sensations measuring environmental noise. J Sound Vib 241:3–18
Ando Y (2002) Correlation factors describing primary and spatial sensations of sound fields. J Sound Vib 258:405–417
Ando Y, Pompoli R (2002) Factors to be measured of environmental noise and its subjective responses based on the model of the auditory-brain system. J Temporal Des Archit Environ 2:2–12. http://www.jtdweb.org/journal/
Ando Y, Saifuddin K, Sato S (2002) Duration sensation when listening to bandpass noises. J Sound Vib 250:31–40
Ando Y (2003) Investigations on cerebral hemisphere activities related to subjective preference of the sound field, published for 1983–2003. J Temporal Des Archit Environ 3:2–27. http://www.jtdweb.org/journal/
Ando Y (2004) On the temporal design of environments. J Temporal Des Archit Environ 4:2–14. http://www.jtdweb.org/journal/
Ando Y (1983) Calculation of subjective preference at each seat in a concert hall. J Acoust Soc Am 74:873–887
Ando Y (1985) Concert hall acoustics. Springer, Heidelberg (Forwarded by Schroeder MR)
Ando Y, Tsuruta H, Motokawa A, Matsushita T, Saifuddin K (1999) Subjective duration of every three-year period for 3 to 18 years of age, estimated by students. J Hum Ergol 28:33–37
Ando Y (2004) On the temporal design of environments. J Temporal Des Arch Environ 4:2–14
Ando Y (2006) Reviews on temporal design for three stages of human life. Most unlikely "time is money", but "time is life". J Temporal Des Arch Environ 6:2–17
Ando Y (2007) Concert hall acoustics based on subjective preference theory. In: Rossing T (ed) Springer handbook of acoustics. Springer, New York, Chap. 10
Ando Y (2009a) Cariani P (ed) Auditory and visual sensations, Springer-Verlag, New York

Ando Y (2009b) Theory of temporal and spatial environmental design. In: McGraw-Hill yearbook of science & technology 2009. McGraw-Hill, New York

Ando Y (2009c) Concert hall acoustics and musical expressions. ARTES, Tokyo (in Japanese)

Ando Y, Ando T (2010) Model of "unconscious" duration experience while listen to music and noise. J Temporal Des Archit Environ 10:1–6. http://www.jtdweb.org/journal/

Ando Y (2011) Brain oriented acoustics. Itto-Sha, Tokyo (in Japanese)

Ando Y (2013) Environmental design for the third stage of human life (persistence of individual creations). J Temporal Des Archit Environ 12:1–12

Ando Y, Cariani P (2014) Neurally based acoustics and visual design. In: Xiang N, Sessler G (eds) Acoustics, information, and communication. Memorial volume in honor of Manfred R. Schroder, Chap. 9

Ando Y (2015) Autocorrelation-based features for speech representation. Acustica united with acustics (in print)

ANSI (1985) Specification for a manikin for simulated in-situ airbone acoustic measurement. Acoustical Society of America, Woodbury

Aoshima N (1981) Computer-generated pulse signal applied for sound measurement. J Acoust Soc Am 69:1484–1488

Ball K, Sekuler R (1979) Masking of motion by broad-band and filtered directional noise. Percept Psychophys 26:206–214

Barron M (1993) Auditorium acoustics and architectural design. E & FN Spon, London

Beranek LL (1962) Music, acoustics, and architecture. Wiley, New York

Bolivar VJ, Cohen AJ, Fentress JC (1994) Semantic and formal congruency in music and motion pictures: effects on the interpretation of visual action. Psychomusicology 13:28–59

Born M, Wolf E (1970) Principles of optics, 4th edn. Pergamon Press, Oxford

Botte MC, Bujas Z, Chocholle R (1975) Comparison between the growth of averaged electroencepharic response and direct loudness estimations. J Acoust Soc Am 58:208–213

Bowen RW, Pokorny J, Smith VC (1989) Sawtooth contrast sensitivity: decrements have the edge. Vis Res 29:1501–1509

Bowen RW, Pokorny J, Smith VC, Fowler MA (1992) Sawtooth contrast sensitivity: effects of mean illuminance and low temporal frequencies. Vis Res 32:1239–1247

Burd AN (1969) Nachhallfreir Musik fuer akustische Modelluntersuchungen. Rundfunktechn. Mitteilungen 13:200–201

Cariani PA, Delgutte B (1996) Neural correlates of the pitch of complex tones. I. Pitch and pitch salience. J Neurophysiol 76:1698–1716

Cariani PA, Delgutte B (1996) Neural correlates of the pitch of complex tones. II. Pitch shift, pitch ambiguity, phase-invariance, pitch circularity, and the dominance region for pitch. J Neurophysiol 76:1717–1734

Cariani P (2001) Temporal coding of sensory information in the brain. Acoust Sci Technol 22:77–84

Chen C, Ando Y (1996) On the relationship between the autocorrelation function of the a-waves on the left and right cerebral hemispheres and subjective preference for the reverberation time of music sound field. J Archit Plann Environ Eng Archit Inst Jpn 489:73–80

Chen C, Ryugo H, Ando Y (1997) Relationship between subjective preference and the autocorrelation function of left and right cortical a-waves responding to the noise-burst tempo. J Archit Plann Environ Eng 497:67–74

Cocchi A, Farina A, Rocco L (1990) Reliability of scale-model researches: a concert hall case. Appl Acoust 30:1–13

Cocchi A (2013) Theatre design in Ancient Times: science or opportunity? Acta Acustica Unit Acustica 99:14–20

Damaske P, Ando Y (1972) Interaural cross-correlation for multichannel loudspeaker reproduction. Acustica 27:232–238

Davies WDT (1966) Generation and properties of maximum-length sequences. Control 10:302–433

Dong DW, Atick JJ (1995) Statistics of natural time-varying images. Network 5:517–548 de Lange H (1952) Experiments on flicker and some calculations on an electrical analogue of the foveal system. Physica 8:935–950

Doi S, Otuka T, Takahashi H (1997) Experimental investigation on lighting control with 1/f fluctuation. IEEJ Trans Electron Inf Syst 117C:409–415 (in Japanese)

Eisner A (1995) Suppression of flicker response with increasing test illuminance: roles of temporal waveform, modulation depth, and frequency. J Opt Soc Am A 12:214–224

Field DJ (1987) Relations between the statistics of natural images and the response properties of cortical cells. J Opt Soc Am A 4:2379–2394

Fraisse P (1984) Perception and estimation of time. Ann Rev Psychol 35:1–36

Gade AC (1989) Investigations of musicians' room acoustic conditions in concert halls. Part I: methods and laboratory experiments. Acustica 69:193–203

Gebhard JW, Mowbray GH (1959) On discriminating the rate of visual flicker and auditory flutter. Am J Psychol 72:521–529

Gros BL, Blake R, Hiris E (1998) Anisotropies in visual motion perception: a flesh look. J Opt Soc Am A 15:2003–2011

Gullikson H (1956) A least square solution for paired comparisons with incomplete data. Psychometrika 21:125–134

Hase S, Takatsu A, Sato S, Sakai H, Ando Y (2000) Reverberance of an existing hall in relation to subsequent reverberation time and SPL. J Sound Vib 232:149–155

Hase S (2001) Reverberance and its control in relation to the physical factors of sound fields in halls. Doctorate dissertation, Graduate School of Science and Technology, Kobe University

Hayashi C (1952) On the prediction of phenomena from qualitative data and the quantification of qualitative data from the mathematico-statistical point of view. Ann Inst Stat Math III:69–98

Hayashi C (1954) Multidimensional quantification. I Proc Jpn Acad 30:61–65

Hayashi C (1954) Multidimensional quantification. II Proc Jpn Acad 30:165–169

Henning GB, Herz BG, Broadbent DE (1975) Some experiments bearing on the hypothesis that the visual system analyzes spatial patterns in independent bands of spatial frequency. Vis Res 15:887–897

Hidaka T (1996) Personal communication

Hidaka T, Beranek LL, Okano T (1995) Interaural cross-correlation, lateral fraction, and low- and high-frequency sound levels as measures of acoustical quality in concert halls. J Acoust Soc Am 98:988–1007

Hidaka T, Beranek L (2000) Objective and subjective evaluations of twenty-three opera house in Europe, Japan an the Americas. J Acoust Soc Am 107:368–383

Holland JH (1975) Adaptation in natural and artificial systems. The University of Michigan Press, Ann Arbor

Houtgast T, Steeneken HJM, Plomp R (1980) Predicting speech intelligibility in rooms from the modulation transfer function, I. General room acoustics. Acustica 46:60–72

Inagaki T, Iizuka K, Agu M, Akabane H, Abe N (2001) $1/f^n$ fluctuating phenomena in luminous pattern of firefly and its healing effect. Trans Jpn Soc Mech Eng 67C:365–372 (in Japanese)

Inoue M, Ando Y, Taguti T (2001) The frequency range applicable to pitch identification based upon the auto-correlation function model. J Sound Vib 241:105–116

Iwamiya S (1994) Interaction between auditory and visual processing when listening to music in an audio visual context: 1. Matching 2. Audio quality Psychomusicology 13:133–154

Jordan VL (1969) Acoustical criteria for auditoriums and their relation to model techniques. J Acoust Soc Am 47:408–412

Kang SH, Ando Y (1985) Comparison between subjective preference judgments for sound fields by different nations. Memoirs Grad Sch Sci Technol Kobe Univ 3-A:71–76

Kaplan S (1987) Aesthetics, affect and cognition: environmental preference from an evolutionary perspective. Environ Behav 19:3–32

Kato K, Ando Y (2002) A study on the blending of vocal music with the sound field by different singing styles. J Sound Vib 258:463–472

Kato K, Fujii K, Kawai K, Ando Y, Yano T (2004) Blending vocal music with the sound field—the effective duration of the autocorrelation function of Western professional singing voices with different vowels and pitches. In: Proceedings of the international symposium on musical acoustics, ISMA 2004, Nara

Kato K, Fujii K, Hirawa T, Kawai K, Yano T, Ando Y (2007) Investigation of the relation between minimum effective duration of running autocorrelation function and operatic singing with different interpretation styles. Acta Acustica Unit Acustica 93:421–434

Katsuki Y, Sumi T, Uchida H, Watanabe T (1958) Electric responses of auditory neurons in cat to sound stimulation. J Neurophysiol 21:569–588

Keet MV (1968) The influence of early lateral reflections on the spatial impression. In: Proceedings of the 6th international congresses on acoustics, Tokyo, Paper E-2-4

Kelly DH (1961) Visual responses to time-dependent stimuli: I. Amplitude sensitivity measurements. J Opt Soc Am 51:422–429

Kiang NY-S (1965) Discharge pattern of single fibers in the cat's auditory nerve. MIT Press, Cambridge, MA

Kimura D (1973) The asymmetry of the human brain. Sci Am 228:70–78

Kinchla RA, Allan LG (1970) Visual movement perception: a comparison of sensitivity to vertical and horizontal movement. Percept Psychophys 8:399–405

Kirkeby O, Nelson PA, Hamada H (1998) The "stereo dipole"—a virtual source imaging system using two closely spaced loudspeakers. J Audio Eng Soc 46:387–395

Korenaga Y (1997) A new method of calculating speech intelligibility with respect to the delay time of reflections. In: Ando Y, Noson D (eds) Conference proceedings of MCHA 1995, Academic Press, London, Chap. 28

Kremers J, Lee BB, Smith VC (1993) Responses of macaque ganglion cells and human observers to compound periodic waveforms. Vis Res 33:1997–2011

Kuttruff H (1991) Room acoustics, 3rd edn. Elsevier Applied Science, London

Levinson E, Sekuler R (1980) A two-dimensional analysis of direction-specific adaptation. Vis Res 20:103–108

Lipscomb SD, Kendall RA (1994) Perceptual judgement of the relationship between musical and visual components in film. Psychomusicology 13:60–98

Marshall AH, Gottlob D, Alrutz H (1978) Acoustical conditions preferred for ensemble. J Acoust Soc Am 64:1437–1442

Marshall AH, Mayer J (1985) The directivity and auditory impressions of singers. Acustica 58:130–140

Mandler MB, Makous W (1984) A three channel model of temporal frequency perception. Vis Res 24:1881–1887

Martignon P, Azzali A, Cabrera D, Capra A, Farina A (2005) Reproduction of auditorium spatial impression with binaural and stereophonic sound system. In: Audio engineering society, 118th convention, Barcelona

Marui A, Martens WL (2005) Constructing individual and group timbre space for sharpness-matched distorted guitar timbres. In: Audio engineering society convention paper, presented at the 119th convention, New York

Meyer J (1995) Influence of communication on stage on the musical quality. In: Proceedings of the 15th international congress on acoustics, Trondheim, pp 573–576

Milizia F (1773) About the Theatre (in Italian), Venezia

Milizia F (1794) A complete Treatise, formal and material, about the Theatre (in Italian), Venezia

Morimoto M, Ando Y (1980) On the simulation of sound localization. J Acoust Soc Jpn 1:167–174

Mosteller F (1951) Remarks on the method of paired comparisons. III. Psychometrika 16:207–218

Mouri K, Akiyama K, Ando Y (2000) Relationship between subjective preference and the alpha-brain wave in relation to the initial time delay gap with vocal music. J Sound Vib 232:139–147

Mouri K, Akiyama K, Ando Y (2001) Preliminary study on recommended time duration of source signals to be analyzed, in relation to its effective duration of autocorrelation function. J Sound Vib 241:87–95

Mouri K, Fujii K, Shimokura R, Ando Y A study on the dynamic properties of auditory interaural cross-correlation function relating to the moving sound image in the horizontal plane for the band noise (Unpublished)

Myers AK, Cotton B, Hilp HA (1981) Matching the rate of concurrent tone bursts and light flashes as a function of surround luminance. Percept Psychophys 30:33–38

Nachmias J, Rogowitz BE (1983) Masking by spatially-modulated gratings. Vis Res 23:1621–1629

Nakajima T (1992) Speech intelligibility and clarity related to spatial-binaural factor for sound field in a room. Ph.D. thesis at Graduate School of Science and Technology, Kobe University

Nakajima T, Ando Y, Fujita K (1992) Lateral low-frequency components of reflected sound from a canopy complex comprising triangular plates in concert halls. J Acoust Soc Am 92:1443–1451

Nakajima T, Yoshida J, Ando Y (1993) A simple method of calculating the interaural cross-correlation function for a sound field. J Acoust Soc Am 93:885–891

Nakayama I (1984) Preferred time delay of a single reflection for performers. Acustica 54:217–221

Nakayama I, Uehata T (1988) Preferred direction of a single reflection for a performer. Acustica 65:205–208

Nakajima T, Ando Y (1997) Calculation and measurement of acoustic factors at each seat in the Kirishima international concert hall. In: Ando Y, Noson D (eds) Music and concert hall acoustics, conference proceedings from MCHA 1995. Chap. 5, pp. 39–49

Noson D, Sato S, Sakai H, Ando Y (2000) Singer responses to sound fields with a simulated reflection. J Sound Vib 232:39–51

Noson D, Sato S, Sakai H, Ando Y (2002) Melisma singing and preferred stage acoustics for singers. J Sound Vib 258:473–485

Okamoto Y, Soeta Y, Ando Y (2003) Analysis of EEG relating to subjective preference of visual motion stimuli. J Temporal Des Archit Environ 3:36–42. http://www.jtdweb.org/journal/

Okamoto Y, Nakagawa S, Yano T, Ando Y (2006) MEG study of cortical responses in relation to subjective preference for the regularity of a fluctuating light. J Temporal Des Archit Environ Published. http://www.jtdweb.org/journal/

Osaki S, Ando Y (1983) A fast method of analyzing the acoustical parameters for sound fields in existing auditoria. In: Proceedings of 4th computer for environmental engineering related to buildings, Tokyo, pp 441–445

Parati L, Pompoli R, Prodi N (2004) The control of balance between singer on the pit and orchestra in the pit by means of virtual opera house models. J Acoust Soc Am 115:2437

Palomaki K, Tiitinen H, Makinen V, May P, Alku P (2002) Cortical processing of speech sounds and their analogues in a spatial auditory environment. Cogn Brain Res 14:294–299

Pompoli R, Prodi N (2000) Guidelines for acoustical measurements inside historical opera houses: procedures and validation. J Sound Vib 232:281–301

Prime E (1994) Measurements of the vibrato rate of ten subjects. J Acoust Soc Am 96:1979–1984

Prime E (1997) Vibrato extent and intonation in professional Western Iyric singing. J Acoust Soc Am 102:616–621

Prodi N, Velecka S (2005) A scale value for the balance inside a historical opera house. J Acoust Soc Am 117:771–779

Raymond JE (1994) Directional anisotropy of motion sensitivity across the visual field. Vis Res 34:1029–1038

Rindel JH (1986) Attenuation of sound reflections due to diffraction. In: Nordic acoustical meeting, pp 20–22

Runderman DL, Bialek W (1994) Statistics of natural images: scaling in the woods. Phys Rev Lett 73:814–817

Sabine WC (1900) Reverberation (the American architect and the engineering record). Prefaced by Beranek LL: Collected papers on acoustics, Peninsula Publishing, Los Altos, California, Chap. 1

van der Schaaf A, van Hateren JH (1996) Modelling of the power spectra of natural images: statistics and information. Vis Res 36:2759–2770

Saifuddin K, Matsushima T, Ando Y (2002) Duration sensation when listening to pure tone and complex tone. J Temporal Des Archit Environ 2:42–47. http://www.jtdweb.org/journal/

Sakai H, Singh PK, Ando Y (1997) Inter-individual differences in subjective preference judgments of sound fields. In: Ando Y, Noson D (eds) Music and concert hall acoustics, conference proceedings of MCHA 1995. Academic Press, London, Chap. 13

Sakai H, Ando Y, Setoguchi H (2000) Individual subjective preference of listeners to vocal music sources in relation to the subsequent reverberation time of sound fields. J Sound Vib 232:157–169

Sakai H, Ando Y, Prodi N, Pompoli R (2002) Temporal and spatial acoustic factors for listeners in the boxes of historical opera theatre. J Sound Vib 258:527–547

Sakai H, Sato S, Prodi N (2004) Orthogonal factors for the stage and pit inside a historical opera house. Acustaica/Acta Acustica 90:319–334

Sakurai M, Aizawa S, Suzumura Y, Ando Y (2000) A diagnostic system measuring orthogonal factors of sound fields in a scale model of auditorium. J Sound Vib 232:231–237

Sato S, Ando Y (1996) Effects of interaural cross-correlation function on subjective attributes. J Acoust Soc Am 100(A):2592

Sato S, Mori Y, Ando Y (1997) On the subjective evaluation of source locations on the stage by listeners. In: Ando Y, Noson D (eds) Music and concert hall acoustics. Academic Press, London, Chap. 12

Sato S, Ando Y (1999) On the apparent source width (ASW) for bandpass noises related to the IACC and the width of the interaural cross-correlation function (W_{IACC}). J Acoust Soc Am 105:1234

Sato S, Ohta S, Ando Y (2000) Subjective preference of cellists for the delay time of a single reflection in a performance. J Sound Vib 232:27–37

Sato S, Ando Y, Mellert V (2001) Cues for localization in the median plane extracted from the autocorrelation function. J Sound Vib 241:53–56

Sato S, Ando Y (2002) Apparent source width (ASW) of complex noises in relation to the interaural cross-correlation function. J Temporal Des Archit Environ 2:29–32. http://www.jtdweb.org/journal/

Sato S, Kitamura T, Ando Y (2002a) Loudness of sharply (2068 dB/Octave) filtered noises in relation to the factors extracted from the autocorrelation function. J Sound Vib 250:47–52

Sato S, Sakai H, Prodi N (2002b) Subjective preference for sound sources located on the stage and in the orchestra pit of an opera house. J Sound Vib 258:549–561

Sato S, Sakai H, Prodi N (2002c) Acoustical measurements in ancient Greek and Roman theatres. In: Proceedings of forum fcusticum 2002: the 3rd European acoustics association convention, Sevilla

Sato S, Otori K, Takizawa A, Sakai H, Ando Y, Kawamura H (2002d) Applying genetic algorithms to the optimum design of a concert hall. J Sound Vib 258:517–526

Sato S, Nishio K, Ando Y (2003) Propagation of alpha waves corresponding to subjective preference from the right hemisphere to the left with change in the IACC of a sound field. J Temporal Des Archit Environ 3:60–69

Sato S, Hayashi T, Takizawa A, Tani A, Kawamura H, Ando Y (2004) Acoustic design of theatres applying genetic algorithms. J Temporal Des Archit Environ 4:41–51

Sato S, Prodi N (2009) On the subjective evaluation of the perceived balance between a singer and a piano inside different theatres. Acta Acustica Unit Acustica 95:519–526

Sakai H, Ando Y, Prodi N, Pompoli R (2002) Temporal and spatial acoustical factors for listeners in the boxes of historical opera theatres

Sakai H, Sato S, Prodi N (2004) Orthogonal factors for the stage and pit inside a historical opera house. ACUSTICA acta acustica 90:319–334

Schroeder MR (1962) Natural sounding artificial reverberation. J Audio Eng Soc 10:219–223

Schroeder MR (1965a) New method of measuring reverberation time. J Acoust Soc Am 37:409–412

Schroeder MR (1965b) Response to “Comments on ‘New method of measuring reverberation time’”. [Smith PW (1965) J Acoust Soc Am 38:359(L)]. J Acoust Soc Am 38:359–361

Schroeder MR, Gottlob D, Siebrasse KF (1974) Comparative study of European concert halls: correlation of subjective preference with geometric and acoustic parameters. J Acoust Soc Am 56:1195–1201

Schroeder MR (1979) Binaural dissimilarity and optimum ceilings for concert halls: More lateral sound diffusion. J Acoust Soc Am 65:958–963

Secker-Walker HE, Searle CL (1990) Time domain analysis of auditory-nerve- fiber firing rates. J Acoust Soc Am 88:1427–1436

Shankland RS (1973) Physics today, October. Acoustics of Greek Theatres

Singh PK, Ando Y, Kurihara Y (1994) Individual subjective diffuseness responses of filtered noise sound fields. Acustica 80:471–477

Soeta Y, Ando Y (2001) Autocorrelation analysis and subjective preferences of images of camphor leaves moving in the wind. J Temporal Des Archit Environ 1:6–11

Soeta Y, Uchida Y, Ando Y (2001) Matching a tonal tempo with camphor leaves moving in the wind. J Temporal Des Archit Environ 1:21–26. http://www.jtdweb.org/journal/

Soeta Y, Okamoto Y, Nakagawa S, Tonoike M, Ando Y (2002a) Autocorrelation analyses of MEG alpha waves in relation to subjective preference of a flickering light. NeuroReport 13:527–533

Soeta Y, Nakagawa S, Tonoike M, Ando Y (2002b) Magnetoencephalographic responses corresponding to individual subjective preference of sound fields. J Sound Vib 258:419–428

Soeta Y, Uetani S, Ando Y (2002c) Relationship between subjective preference and alpha wave activity in relation to temporal frequency and mean luminance of a flickering light. J Opt Soc Am A 19:289–294

Soeta Y, Nakagawa S, Tonoike M, Ando Y (2003a). Spatial analysis of magnetoencephalographic alpha waves in relation to subjective preference of a sound field, J Temporal Des Archit Environ 3:28–35. http://www.jtdweb.org/journal/

Soeta Y, Ohtori K, Ando Y (2003b) Subjective preference for movements of a visual circular stimulus: a case of sinusoidal movement in vertical and horizontal directions. J Temporal Des Archit Environ 3:70–76. http://www.jtdweb.org/journal/

Soeta Y, Nakagawa S, Tonoike M (2005a) Magnetoencephalographic activities related to the magnitude of the interaural cross-correlation function (IACC) of sound fields. J Temporal Des Archit Environ 5:5–11. http://www.jtdweb.org/journal/

Soeta Y, Mizuma K, Okamoto Y, Ando Y (2005b) Effects of the degree of fluctuation on subjective preference for a 1 Hz flickering light. Perception 34:587–593

Soeta Y, Nakagawa S (2006) Auditory evoked magnetic fields in relation to interaural time delay and interaural crosscorrelation. Hear Res 220:106–115

Soeta Y, Ando Y (2015) Neurally-based measurement and evaluation of environmental noise. Springer, Tokyo (to be published)

Sperry RW (1974) Lateral specialization in the surgically separated hemispheres. In: Schmitt FO, Worden FC (eds) The neurosciences: third study program, MIT Press, Cambridge, Chap. 1

Steinhardt PJ (2009) Cyclic universe theory. In: McGraw-Hill yearbook of science & technology. McGraw-Hill, New York, pp 69–71

Sugano Y, Iwamiya S (1999) Effects of synchronization between musical rhythm and visual motion on the congruency of music and motion picture (in Japanese). J Music Percept Cogn 5:1–10

Sugano Y, Iwamiya S (2000) The effects of synchronization between auditory and visual accents and those of matching between musical tempo and visual speed on the emotional impression of combinations of motion picture and music (in Japanese). J Acoust Soc Jpn 56:695–704

Sumioka T, Ando Y (1996) On the pitch identification of the complex tone by the autocorrelation function (ACF) model. J Acoust Soc Am 100(A):2720

Suzumura Y, Sakurai M, Yamamoto I, Iizuka T, Oowaki M, Ando Y (2000) An evaluation of the effects of scattered reflections in a sound fields. J Sound Vib 232:303–308

Taguti T, Ando Y (1997) Characteristics of the short-term autocorrelation function of sound signals in piano performances. In: Ando Y, Noson D (eds) Music and concert hall acoustics, conference proceedings of MCHA 1995, Academic Press, London, Chap. 23

Takatsu A, Hase S, Sakai H, Sato S, Ando Y (2000) Acoustical design and measurement of a circular hall, improving both spatial and temporal factors at each seat. J Sound Vib 232:263–273

Thurstone LL (1927) A law of comparative judgment. Psychol Rev 34:273–289

Torgerson WS (1958) Theory and methods of scaling. Wiley, New York

Tronchin L, Farina A (1997) Acoustics of the former teatro “La Fenice” in Venice. J Audio Eng Soc 45:1051–1062

van de Grind WA, Koenderink JJ, van Doorn AJ, Milders MV, Voerman H (1993) Inhomogeneity and anisotropies for motion detection in the monocular visual field of human observers. Vis Res 33:1089–1107

Vitruvius (ca 25 BC) (1960) De architecture, Liber V, Cap. VIII. (de locis consonantibus ad theatra eligendis). The ten books on architecture (trans: Morgan MH). Dover, New York

Voss RF, Clarke J (1978a) “1/f noise” in music and speech. Nature 258:317–318

Voss RF, Clarke J (1978b) “1/f noise” in music: music from 1/f noise. J Acoust Soc Am 63:258–263

Wu S, Burns SA, Reeves A, Elsner AE (1996) Flicker brightness enhancement and visual nonlinearity. Vis Res 36:1573–1583

Yost WA (1996) A time domain description for the pitch strength of iterated rippled noise. J Acoust Soc Am 99:1066–1078

Zwicker E, Fastl H (1999) Psychoacoustics. Springer, New York

Index

Y. Ando, *Opera House Acoustics Based on Subjective Preference Theory*,
Mathematics for Industry 12, DOI 10.1007/978-4-431-55423-3

MIX
Papier aus verantwortungsvollen Quellen
Paper from responsible sources
FSC® C105338

If you have any concerns about our products,
you can contact us on
ProductSafety@springernature.com

In case Publisher is established outside the EU,
the EU authorized representative is:
Springer Nature Customer Service Center GmbH
Europaplatz 3, 69115 Heidelberg, Germany

Printed by Libri Plureos GmbH
in Hamburg, Germany